Natural Environment Research Council

Cotton strip assay: an index of decomposition in soils

(ITE symposium no. 24)

Edited by
A F HARRISON, P M LATTER and D W H WALTON

INSTITUTE OF TERRESTRIAL ECOLOGY

Printed in Great Britain by Titus Wilson & Son Ltd.
© NERC Copyright 1988

Published in 1988 by
Institute of Terrestrial Ecology
Merlewood Research Station
GRANGE-OVER-SANDS
Cumbria
LA11 6JU

BRITISH LIBRARY CATALOGUING-IN-PUBLICATION DATA
Cotton strip assay: an index of decomposition in soils
    ITE symposium, ISSN 0263-8614; no. 24)
    1. Cellulose — Biodegradation — Measurement
    2. Soils — Analysis
    I. Harrison, A.F. II. Latter, P.M. III. Walton, D.W.H.
    IV. Institute of Terrestrial Ecology
    631.4'17   S592.6.C4
    ISBN 1 870393 06 6

COVER ILLUSTRATION
Designed by C. B. Benefield

The *Institute of Terrestrial Ecology (ITE)* was established in 1973, from the former Nature Conservancy's research stations and staff, joined later by the Institute of Tree Biology and the Culture Centre of Algae and Protozoa. ITE contributes to, and draws upon, the collective knowledge of the 14 sister institutes which make up the *Natural Environment Research Council,* spanning all the environmental sciences.

The Institute studies the factors determining the structure, composition and processes of land and freshwater systems, and of individual plant and animal species. It is developing a sounder scientific basis for predicting and modelling environmental trends arising from natural or man-made change. The results of this research are available to those responsible for the protection, management and wise use of our natural resources.

One quarter of ITE's work is research commissioned by customers, such as the Department of Environment, the European Economic Community, the Nature Conservancy Council and the Overseas Development Administration. The remainder is fundamental research supported by NERC.

ITE's expertise is widely used by international organizations in overseas projects and programmes of research.

Dr A F Harrison,
Miss P M Latter   and Dr D W H Walton
Institute of Terrestrial Ecology   British Antarctic Survey
Merlewood Research Station   High Cross
GRANGE-OVER-SANDS   Madingley Road
Cumbria   CAMBRIDGE
LA11 6JU   CB3 OET
044 84 (Grange-over-Sands) 2264

CONTENTS

4

FIELD APPLICATION IN SPECIFIC ENVIRONMENTS

Temperate

Tropical

Polar

International comparisons

APPENDICES

EPILOGUE

PREFACE

Nutrient cycling in natural, silvicultural forest or agricultural ecosystems is maintained by continued decomposition of organic matter on and within the soil. The rate of organic matter decomposition is a complex function of litter quality, soil and environmental factors and, despite considerable research, it has not proved easy to quantify or to model the interactions involved.

To investigate either the direct or indirect effects of soil and environmental factors separately from those of litter quality, many research workers have employed standardized organic substrates, as analogues in decomposition studies. One such substrate is cotton fabric strips, a robust material but readily conforming to soil shape. It is usually inserted vertically into soil for a period of time, with their degree of decomposition being assessed by the loss in tensile strength of the cotton.

The standardized procedure which has been developed gives an integrated result over a particular period of time, and is here referred to as the 'cotton strip assay'. Because the assay reacts to the entirety of changes in environmental factors which control decomposition (temperature, moisture, nutrient availability), it is used where a comparative measure is required for the effects of natural variations or of management practices. Although initially used as a comparative index of cellulose decomposition in surveys, it is now frequently considered as a general index of decomposer potential, as the results in this publication demonstrate.

The assay has been used in many types of environment with a wide geographical spread, with modifications for experimental studies on macro- and microscales. This diversity of uses has led to many publications throughout the scientific literature. It was, therefore, considered opportune to bring together those using the assay at a Symposium in Grange-over-Sands, which was held in October 1985, to discuss techniques, to pool available data, and to consider further possible developments of the assay.

This publication presents the papers and topics discussed. It is intended not only as a report of the meeting, but more as a handbook and reference base for the cotton strip assay. For this reason, the papers were revised after the meeting following the discussions which took place, and have been fully refereed and edited. The editors are most grateful to the various referees and, in particular, to Professor A Macfadyen and Dr J C Frankland for their valued contribution. We also acknowledge the considerable assistance from typists, graphics programmers and many others who have helped in numerous ways.

A F Harrison
P M Latter

GENERAL METHOD AND THEORY

The cotton strip assay for cellulose decomposition studies in soil: history of the assay and development

P M LATTER[1] and D W H WALTON[2]
[1]Institute of Terrestrial Ecology, Merlewood Research Station, Grange-over-Sands
[2]British Antarctic Survey, Cambridge

1 Summary
The history and development of the cotton strip assay as a field method for ecological use in the International Biological Programme (IBP) and other studies are discussed. The further standardization and progressive refinement of the assay have reduced many of the original uncertainties regarding the data. Use of the cotton strip assay in ecological studies is described.

2 History of soil burial methods
The textile industry has long been concerned with the prevention of cloth decay, especially by fungal attack. To test the effectiveness of fungicide treatments, a routine test has been devised in which strips of textile are buried in tanks of soil (Wade 1947; Barr 1988; etc). The loss in tensile strength of the cotton fibres, with time, is used as a quantitative assessment of the rottenness of the fabric and, thus, the effectiveness of the fungicide. The value of the test and its widespread use stimulated the specification of the British Standard 2576 (Anon 1986) for general commercial application of the test.

As the soil burial test became more widely used, the variability inherent in the use of different types of cloth and soil became clear (Schmidt & Ruschmeyer 1958). One way of reducing variation inherent in the test was to standardize the cloth type used. Restriction of cloth type to one general group, for instance unbleached calico, could, however, only be a partial solution, as different fibre mixtures, thread densities and weaving treatments in different mills could produce wide variations in tensile strength (Sagar 1988a). A specialized cloth was, therefore, required and in the 1960s Toegepast Natuurwetenschappelijk Onderzoek (TNO) in the Netherlands produced such a material.

The breakdown of cotton fibres is of interest to a range of textile scientists, and early work is summarized in Thaysen and Bunker (1927) and Siu (1951). Research has been carried out on the soil burial method by the textile industry, to develop an understanding of the test (Barr 1988) and the optimum conditions for microbial attack. Much of this research used cotton, and was carried out at the Shirley Institute, Manchester. Detailed studies were made of the method of attack, both physical (Simpson & Marsh 1960; Kassenbeck 1970) and chemical aspects (Blum & Stahl 1952; Selby 1961; Halliwell 1965; Sternberg et al. 1977) and the

cellulolytic enzymes involved (Sagar 1988b). A range of information was thus available by the early 1960s about a test which measured the effects of biological attack on a partially processed natural substrate which was generally over 96% pure cellulose.

The decomposition of cotton in soil appears largely to be due to microbial attack (Latter et al. 1988), a conclusion to be confirmed for normal soil temperatures by Richard (1945) during storage of cotton at pH 2.

3 Development of the cotton strip assay for soil studies
Cellulose comprises the bulk of plant material and is of great significance in the decomposition of soil organic matter by micro-organisms and microfauna, and thus in nutrient cycling, in natural and agricultural ecosystems. Although cellophane film, filter paper and lens tissue have been successfully used in cellulose decomposition experiments, the need for a more robust, flexible, standard substrate for ecological studies in soil was recognized many years ago. As early as 1945, Richard (1945) used cellulose cords of viscose rayon (regenerated cellulose fibre) in alpine areas of Switzerland, whilst in 1948, in Poland, Kuzniar (1948, 1988) used strips of linen (a partially lignified fibre), as do other east European workers still (Strzelczyk et al. 1978; Sadanov 1982). Cotton tape was also used in Australian soil studies (Rovira 1953) and for assessing cellulase activity of isolates of the fungi Pythium (Taylor & Marsh 1963) and Chaetomium (Farrow 1951).

Following on from Kuzniar's experiment and the soil burial method, strips of cotton (unbleached calico) with over 95% cellulose were inserted vertically into soil profiles (Plate 6) for cellulose decomposition studies on Pennine moorland sites (Latter et al. 1967). Visual observations showed marked differences in rotting of the cloth between sites, and the assay was quantified by using tensile strength as the measure of cloth decomposition. The purpose of the soil burial test was, in essence, reversed to develop a field ecological test (Table 1 & Plate 9).

The development of the International Biological Programme (IBP) from 1965 onwards included investigations of the decomposer cycle in various ecosystems. In the Tundra Biome, it was decided

Table 1. A comparison of the procedures and purpose of the soil burial test as used in the textile industry and the cotton strip assay used in field ecology

Soil burial textile test	Cotton strip assay as soil test
Procedures:	Procedures:
Standard soil	Variable soils
Standard or selected textiles	Standard textile-cotton
Variable treatment of textile	Standard treatment of textile
Frayed before burial	Frayed after burial
Controlled environment, high incubation temperature and moisture	Varying environment in field, or controlled for experiments
Soil is used to show:	Cotton is used to show:
Rotting of textile with varying treatments, eg fungicides, or textile manufacturing variations	Ability of soil to rot cotton with varying soils or site treatments
Suitability of textile for use under particular environmental conditions	Suitability of soil environment for biological activity

to concentrate on the study of energy flow through various trophic levels (Dahl & Gore 1968). With sites spread over the polar and alpine regions of the world, an assay was needed to provide standardized inter-site comparisons of rates of cellulose decomposition in soil. A paper cellulose board (Rosswall 1974) was used along with cotton textile strips (Heal *et al.* 1974), and the latter became the most widely used method providing, for the first time, comparable data on cellulose decay from the arctic, antarctic and alpine sites (Heal *et al.* 1974). It, thus, became possible to test for the effectiveness of a particular subset (the cellulolytic species) of the cosmopolitan soil flora on a global scale.

To provide some degree of standardization, a single source of cotton, an unbleached commercial calico (Plate 7), continued to be supplied to IBP field scientists from stocks at the Institute of Terrestrial Ecology's Merlewood Research Station. Techniques for field use were later refined and standardized (Latter & Howson 1977). One necessary step was the provision of adequate supplies of a closely specified test cloth, as the TNO fabric was no longer available. The Shirley Institute produced the 'Shirley Soil Burial Test Fabric' (Plate 7) in 1976 (Walton & Allsopp 1977; Sagar 1988a). The cloth was widely advertized and was soon in use for commercial testing and for ecological studies.

The assay has now been used, in particular, for polar studies (eg Davis 1986; Wynn-Williams 1988) and for land management studies (eg Brown 1988; Brown & Howson 1988), and it has become apparent that certain aspects, including some procedures taken over from the soil burial test, need further assessment in the context of the ecological test with its high field variability. Some technical aspects, including autoclaving to sterilize cloth prior to burial, washing to clean

retrieved cloth for storage until tensile tested, the best conditions for storage of strips, the effect of width, the need for fraying of substrips for tensile testing, and the inter-comparability of different tensile testing machines, are reported by Latter *et al.* (1988) and Walton (1985). The method for the assay, as we would currently recommend it, is given in Appendix I.

4 Ecological use of the assay

The impetus for more widespread use of the cotton strip assay in soil was undoubtedly provided by the successful world-wide experiments in the IBP Tundra Biome. The development of the Shirley Soil Burial Test Fabric and its incorporation into the British Standard 6085 (Anon 1981) commercial test provided, for the first time, a single source of standard textile for all users. A wide variety of uses has ensued.

Commercial users continue to carry out fungicide assessment tests in laboratory soil tanks (Anon 1981). The Shirley Soil Burial Test Fabric has also been widely used ecologically for field comparisons of cellulose decomposition, having been supplied to research workers in UK, USA, New Zealand, Australia, Canada and the Antarctica. During IBP, the earlier type of cloth was also used in Finland, Ireland, Norway, USSR, Sweden and Alaska. Less orthodox usages have included assessments of cellulose decomposition in coal tips, compost heaps, and the rumen of sheep.

Many of the technical problems originally associated with the cotton strip assay have now been adequately documented. We do not suggest that we understand exactly what occurs during biological attack, as much still remains to be learnt about the biochemistry of cellulose decomposition (Sagar 1988b; Howard 1988). Nevertheless, the technique and data analysis now appear sufficiently standardized for users to have confidence in the inter-comparability of data.

The papers in this volume emphasize the practical nature of the assay. It is especially useful for early detection of biological change in soils or in pinpointing aspects of environmental change needing more detailed investigation, eg the marked depression of cellulose decomposition on an oak (*Quercus* spp.) site at 4–12 cm depth during summer (Brown 1988), the enhanced cellulose decomposition in sub-antarctic sites in the vicinity of rush (*Juncus* spp.) roots (Smith & Walton 1988), or the unexpectedly high cellulose decomposition at Rothera Point, Antarctica, possibly due to radiant energy input (Wynn-Williams 1988). Information is currently available for a number of sites world-wide, and one purpose of the Workshop was to bring these data together. The Workshop results are discussed by Ineson *et al.* (1988).

Some evidence is available suggesting that cellulose decomposition relates to soil properties similar to those affecting plant growth (Latter & Harrison 1988).

Cotton strip assay data show similar trends to weight loss of plant litter (French 1988), but a close relationship cannot be expected (Howard 1988) because the decomposition of cellulose in plant litter is influenced in various ways by other chemical constituents to which it is bound or closely allied. Partition of decomposition stages, thus, requires the use of individual standard substrates to simplify the system for examining the various influences of environmental parameters. Decomposition of any single substrate can never be fully representative of any other, and this is particularly true of cellulose. However, Shawky and Hickisch (1984) show virtually the same ratio of weight losses (1.1–1.2) when comparing 2 soils for decomposition of cotton, filter paper or wheat straw by *Trichoderma*.

The use of this type of assay in soil, which integrates decomposing activity over a period of time, partly overcomes a sampling problem when working at widely dispersed sites. Short-term changes in environmental conditions during the time taken to sample remote sites should have no significant effect on an integrated measurement, but could produce large differences in organism activity, if instantaneous measurements, such as respiration or enzyme assays (eg cellulase), are used. It is, however, obvious that any decomposition test involving the addition of a substrate to soil is no absolute measure of activity, but only an index of the potential to attack that added substrate. The same applies when natural litter falls or is incorporated into soil.

As well as using the assay to examine cellulose decomposition unconfounded by other factors, which was the original purpose, the assay is often considered useful as an index of general biological, or decomposer, activity, on the basis that cellulose is a major part of organic matter. Both uses are discussed in various papers in this volume. As different advantages and limitations of using the assay apply in the 2 approaches, the aims of any study and the purpose of using the assay should be examined critically from the start.

Many of the criticisms raised by Howard (1988) are valid, and the suggested need for 'a method for studying the decomposition of cellulose, and other constituents, in plant litter' cannot be disputed, but the current absence of a method suitable for field use means that analogue substances will remain, for the present time, a practical means of assessing certain biological changes. It would certainly be advantageous if similar simple field tests could be developed for other substrates, eg chitin sheet or complexes of lignin and cellulose. We have had no success with silk as a protein substrate as used by Richard (1945), as it appeared to be chemically or photochemically attacked, but chitin sheet and wood veneer (Latter 1984) appeared promising. In the meantime, we consider that useful ecological and management information

can be obtained with this cotton strip assay – provided users are well aware, as with any method, of its limitations and that use of a simple method does not imply careless attention to detail.

5 *Acknowledgements*
The authors acknowledge the many people who have contributed to the development of the method, including staff at ITE Merlewood, and also staff at the Shirley Institute for much theoretical and practical advice.

6 *References*
Anon. 1981. *Method of test for the determination of the resistance of textiles to microbiological deterioration.* (BS 6085.) London: British Standards Institution.
Anon. 1986. *Methods of test for textiles - woven fabrics - determination of breaking strength and elongation (strip method).* (BS 2576.) London: British Standards Institution.
Barr, A.R.M. 1988. The colonization and decay of cotton by fungi in soil burial tests used in the textile industry. In: *Cotton strip assay: an index of decomposition in soils,* edited by A.F. Harrison, P.M. Latter & D.W.H. Walton, 50-54. (ITE symposium no. 24.) Grange-over-Sands: Institute of Terrestrial Ecology.
Blum, R. & Stahl, W.H. 1952. Enzyme degradation of cellulose fibres. *Text. Res. J.,* **22,** 178-192.
Brown, A.H.F. 1988. Discrimination between the effects on soils of 4 tree species in pure and mixed stands using cotton strip assay. In: *Cotton strip assay: an index of decomposition in soils,* edited by A.F. Harrison, P.M. Latter & D.W.H. Walton, 80-85. (ITE symposium no. 24.) Grange-over-Sands: Institute of Terrestrial Ecology.
Brown, A.H.F. & Howson, G. 1988. Changes in tensile strength loss of cotton strips with season and soil depth under 4 tree species. In: *Cotton strip assay: an index of decomposition in soils,* edited by A.F. Harrison, P.M. Latter & D.W.H. Walton, 86-89. (ITE symposium no. 24.) Grange-over-Sands: Institute of Terrestrial Ecology.
Dahl, E. & Gore, A.J.P. 1968. *Proceedings of working meeting on analysis of ecosystems: tundra zone.* Utaoset, Norway: International Biological Programme.
Davis, R.C. 1986. Environmental factors influencing decomposition rates in two antarctic moss communities. *Polar Biol.,* **5,** 95-104.
Farrow, W.M. 1951. A study of *Chaetomium* in cellulose decay. *Proc. Iowa Acad. Sci.,* **58,** 101-106.
French, D.D. 1988. Patterns of decomposition assessed by the use of litter bags and cotton strip assay on fertilized and unfertilized heather moor in Scotland. In: *Cotton strip assay: an index of decomposition in soils,* edited by A.F. Harrison, P.M. Latter & D.W.H. Walton, 100-108. (ITE symposium no. 24.) Grange-over-Sands: Institute of Terrestrial Ecology.
Halliwell, G. 1965. Hydrolysis of fibrous cotton and reprecipitated cellulose by cellulolytic enzymes from soil micro-organisms. *Biochem. J.,* **95,** 270-281.
Heal, O.W., Howson, G., French, D.D. & Jeffers, J.N.R. 1974. Decomposition of cotton strips in tundra. In: *Soil organisms and decomposition in tundra,* edited by A.J. Holding, O.W. Heal, S.F. MacLean & P.W. Flanagan, 341-362. Stockholm: Tundra Biome Steering Committee.
Howard, P.J.A. 1988. A critical evaluation of the cotton strip assay. In: *Cotton strip assay: an index of decomposition in soils,* edited by A.F. Harrison, P.M. Latter & D.W.H. Walton, 34-42. (ITE symposium no. 24.) Grange-over-Sands: Institute of Terrestrial Ecology.
Ineson, P., Bacon, P.J. & Lindley, D.K. 1988. Decomposition of cotton strips in soil: analysis of the world data set. In: *Cotton strip assay: an index of decomposition in soils,* edited by A.F. Harrison, P.M. Latter & D.W.H. Walton, 155-165. (ITE symposium no. 24.) Grange-over-Sands: Institute of Terrestrial Ecology.
Kassenbeck, P. 1970. Bilateral structure of cotton fibres as revealed by enzymatic degradation. *Text. Res. J.,* **40,** 330-334.
Kuzniar, K. 1948. (Studies on the cellulose decomposition in forest soil). *Inst. Bad. Lesn. Rozpr. i Spraw,* **50,** 1-44.

Kuzniar, K. 1988. Examination of the biological activity of the soil under natural conditions. In: *Cotton strip assay: an index of decomposition in soils*, edited by A.F. Harrison, P.M. Latter & D.W.H. Walton, 114-116. (ITE symposium no. 24.) Grange-over-Sands: Institute of Terrestrial Ecology.

Latter, P.M. 1984. The use of wood veneer in decomposition experiments. *Appl. Biochem. Biotechnol.*, **9,** 371.

Latter, P.M. & Harrison, A.F. 1988. Decomposition of cellulose in relation to soil properties and plant growth. In: *Cotton strip assay: an index of decomposition in soils*, edited by A.F. Harrison, P.M. Latter & D.W.H. Walton, 68-71. (ITE symposium no. 24.) Grange-over-Sands: Institute of Terrestrial Ecology.

Latter, P.M. & Howson, G. 1977. The use of cotton strips to indicate cellulose decomposition in the field. *Pedobiologia,* **17,** 145-155.

Latter, P.M., Cragg, J.B. & Heal, O.W. 1967. Comparative studies on the microbiology of four moorland soils in the northern Pennines. *J. Ecol.,* **55,** 445-464.

Latter, P.M., Bancroft, G. & Gillespie, J. 1988. Technical aspects of the cotton strip assay method. *Int. Biodeterior,* **24.** In press.

Richard, F. 1945. The biological decomposition of cellulose and protein test cords in soils under forest and grass associations. I. The method of determining biological soil activity by the so-called 'tearing' test. *Mitt. schweiz. Anst. forstl. VersWes.,* **24,** 297-397.

Rosswall, T. 1974. Cellulose decomposition studies on the tundra. In: *Soil organisms and decomposition in tundra*, edited by A.J. Holding, O.W. Heal, S.F. MacLean & P.W. Flanagan, 325-340. Stockholm: Tundra Biome Steering Committee.

Rovira, A.D. 1953. A study of the decomposition of organic matter in red soils of the Lismore district. *Aust. Conf. Soil Sci., Adelaide,* **1,** 3.17, 1-4.

Sadanov, A.K. 1982. Changes in the number of microorganisms and mobile nitrogen level after application of ameliorants to saline flooded soil. *Izv. Akad. Nauk. Kaz. Ssr Ser. Biol.,* (3), 53-56.

Sagar, B.F. 1988a. The Shirley Soil Burial Test Fabric and tensile testing as a measure of biological breakdown of textiles. In: *Cotton strip assay: an index of decomposition in soils*, edited by A.F. Harrison, P.M. Latter & D.W.H. Walton, 11-16. (ITE symposium no. 24.) Grange-over-Sands: Institute of Terrestrial Ecology.

Sagar, B.F. 1988b. Microbial cellulases and their action on cotton fibres. In: *Cotton strip assay: an index of decomposition in soils*, edited by A.F. Harrison, P.M. Latter & D.W.H. Walton, 17-20. (ITE symposium no. 24.) Grange-over-Sands: Institute of Terrestrial Ecology.

Schmidt, E.L. & Ruschmeyer, O.R. 1958. Cellulose decomposition in soil burial beds. I. Soil properties in relation to cellulose degradation. *Appl. Microbiol.,* **6,** 108-114.

Selby, K. 1961. The degradation of cotton cellulose by the extracellular cellulase of *Myrothecium verrucaria. Biochem. J.,* **79,** 562-566.

Shawky, B.T. & Hickisch, B. 1984. Cellulolytic activity of *Trichoderma* sp. strain G, grown on various cellulose substrates. *Zent.bl. Mikrobiol.,* **139,** 91-96.

Simpson, M.E. & Marsh, P.B. 1960. The decomposition of the cellulose of cotton fibres by fungi of the genus *Aspergillus. Develop. ind. Microbiol.,* **1,** 248-252.

Siu, R.G.H. 1951. *Microbial decomposition of cellulose.* New York: Reinhold.

Smith, M.J. & Walton, D.W.H. 1988. Patterns of cellulose decomposition in four subantarctic soils. *Polar Biol.* In press.

Sternberg, D., Vijayakumar, P. & Reese, E.T. 1977. β-glucosidase: microbial production and effect on enzymatic hydrolysis of cellulose. *Can. J. Microbiol.,* **23,** 139-147.

Strzelczyk, E., Stopinski, M. & Dziadowiec, H. 1978. Studies on cellulose decomposition in forest soils of the reserve 'Las Piwnicki' near Torun. *Acta Univ. Nicolai Copernic. Biol.,* **21,** 115-128.

Taylor, E.E. & Marsh, P.B. 1963. Cellulose decomposition by *Pythium. Can. J. Microbiol.,* **9,** 353-358.

Thaysen, A.C. & Bunker, H.J. 1927. *The microbiology of cellulose, hemicelluloses, pectin and gums.* London: Oxford University Press.

Wade, G.C. 1947. Effect of some micro-organisms on the physical properties of cotton duck. *J. Coun. scient. ind. Res. Aust.,* **20,** 459-467.

Walton, D.W.H. 1985. Tensometer and jaw types in testing the tensile strength of textiles. *Int. Biodeterior.,* **21,** 301-302.

Walton, D.W.H. & Allsopp, D. 1977. A new test cloth for soil burial trials and other studies on cellulose decomposition. *Int. Biodeterior. Bull.,* **13,** 112-115.

Wynn-Williams, D.D. 1988. Cotton strip decomposition in relation to environmental factors in the maritime Antarctic. In: *Cotton strip assay: an index of decomposition in soils*, edited by A.F. Harrison, P.M. Latter & D.W.H. Walton, 126-133. (ITE symposium no. 24.) Grange-over-Sands: Institute of Terrestrial Ecology.

The Shirley Soil Burial Test Fabric and tensile testing as a measure of biological breakdown of textiles

B F SAGAR
Shirley Institute, Manchester

1 Summary

The Shirley Soil Burial Test Fabric has been developed, for use by ecologists as well as by the textile industry, as a standard woven cotton fabric for assessing the cellulolytic microbial activity in soils. The plain woven cloth contains combed 100% cotton 2-fold yarns, with coloured marker threads introduced at 5 and 10 mm intervals in the warp to enable specimens to be easily prepared in various widths according to requirements. The fabric is produced to a high standard and is as free as possible from any extraneous matter, with the dyestuff in the marker threads free from any biocidal effects.

The cellulolytic activity in the soil is assessed by determining the reduction in tensile strength of the buried test specimens, but, in the interpretation, it is important to be aware of the various factors affecting the tensile properties of textile materials and the results obtained from different testing instruments.

2 Use of cotton fabric to assess the cellulolytic activity of soil

The cotton strip assay, based on losses in tensile strength of standard strips of cotton fabric placed in soil, is widely used by ecologists as a simple assay of the cellulolytic activity of the soil. Cotton fibres consist of unicellular seed hairs tapering from base to tip, with variable length and diameter depending on the variety of the cotton; typical fibre lengths are about 2.5 cm, with a diameter of about 15 μ. The cotton hair in the unopened boll on the plant is a hollow, thick-walled cylinder (Figure 1), with a lumen containing protoplasm and cell fluids. The exterior surface of the raw cotton fibre is a coherent membrane (the cuticle), consisting of wax and pectic substances. The cuticle is removed by kier boiling to expose the primary wall, which consists of a loose and somewhat felt-like

assembly of criss-crossing microfibrils, each about 200–300 Å thick. The bulk of the cellulose is contained in the secondary wall (lying immediately inside the primary wall) and is a tightly packed and well-oriented assemblage of fibrils. The fibrils seen with the optical microscope are at least 200 Å thick, but these are clearly bundles of much smaller units, and the electron microscope reveals single elementary fibrils with a spread of diameter below 75 Å (ie no wider than the width of a single crystallite, as deduced from X-ray measurements). Aggregation of elementary fibrils into microfibrils (200–300 Å thick) and then into macrofibrils (1000 Å) is evident in electron micrographs (Plate 1).

Plate 1. Microfibrillar structure of cotton seen under the transmission electron microscope

When the boll opens and the fibres dry, they collapse to give a twisted, ribbon-like structure (Plate 2 iv). This collapse of the fibre on drying has been shown by Kassenbeck (1970) to result in a bilateral structure. Four zones can be distinguished in the fibre cross-section, which differ in the organization of their fine structure as a result of the radial and tangential compressive forces produced in the fibre during its initial desiccation. These 4 zones also differ in their accessibility to cellulase enzymes, and those parts of the cotton hair with the more open morphological structure are the first to be attacked, as is seen clearly under the scanning electron microscope (Plate 2 ix–x).

In terms of direct observations on cotton fibres *in vitro*, 6 stages of enzymic attack are observed (Table 1). Loss of tensile strength is a sensitive measure of attack, and the cotton loses strength much faster than it loses weight when repeatedly treated with a crude cell-free enzyme preparation of relatively low enzyme activity (Figure 2).

Measurements of changes in degree of polymerization, increase in solubility in sodium hydroxide (10%),

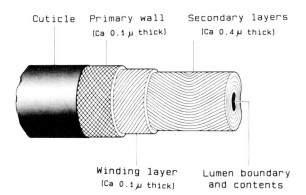

```
 Cuticle   Primary wall      Secondary layers
          (Ca 0.1 μ thick)   (Ca 0.4 μ thick)




      Winding layer       Lumen boundary
      (Ca 0.1 μ thick)      and contents
```

Figure 1. Schematic diagram of layered components of the cotton fibre cell wall

loss in weight, and production of reducing sugars have all been used to quantify the progress of cellulose enzymolysis in the laboratory, but determination of the progressive loss of tensile strength of buried cotton strips is the most practicable way of assessing the cellulolytic activity of soil.

3 The Shirley Soil Burial Test Fabric

The possibility that the Shirley Institute might produce a standard fabric for soil burial tests was first suggested in 1975 by Dr D W H Walton of the Natural Environment Research Council's British Antarctic Survey. A test fabric made by Toegepast Natuurwetenschappelijk Onderzoek (TNO) in the Netherlands ceased to be available, and there was general consensus that a Shirley Soil Burial Test Fabric should be produced which closely resembled the TNO specification. More than 20 organizations expressed their keen interest in the initial development of a suitable test fabric.

A plain woven fabric was produced containing combed 100% cotton yarns in a 2-folded form with the specifications given in Table 2 and as recommended in British Standards 6085 (Anon 1981). Package-dyed yarn was introduced into the warp as marker threads (Plate 7) to enable specimens to be prepared in 10 mm, 20 mm, 25 mm, 30 mm, etc, widths according to requirements. It was obviously important that the dyestuff used should not affect the molecular structure of the cellulose (therefore, reactive dyes were ruled out), neither could they have any biocidal or biochemical influence. A further requirement was that the dyed yarn should be fast to kier boiling, and not 'run' into

Table 1. Six stages of cellulase enzyme attack on cotton fibres

1. Increase in alkali swelling (18% NaOH)
2. Transverse incipient cracking
3. Loss of tensile strength
4. Lowering degree of polymerization (DP)
5. Increase in alkali solubility (10% NaOH)
6. Loss in weight and production of reducing sugars

Table 2. Shirley Soil Burial Test Fabric

The plain woven cloth contains combed 100% cotton yarns (American type cotton, good middling) in a 2-folded form to the following specification
(Linear density of yarn is given as: yarn count, ie number of 840-yd hanks weighing 1lb (Ne), with 's signifying a 2-fold yarn, each component with the given yarn count; weight of yarn in g 1000 m^{-1} (tex) where R indicates that the tex value is for the combined linear density of 2 single yarns; Z and S conventions describe direction of yarn twist)

| | 1978 fabric (Green marker threads)[1] | | 1981 fabric (Blue marker threads)[1] | |
	Warp	Weft	Warp	Weft
Yarn count weight	2/32's Ne R 37.0/2	2/20's Ne R 59.0/2	2/32's Ne R 37.0/2	2/20's Ne R 59.0/2
Singles twist (turns m^{-1})	890 Z	630 Z	885 Z	885 Z
Folded twist (turns m^{-1})	750 S	710 S	748 S	748 S
Threads cm^{-1}	34	17	34	17

[1]Marker threads introduced in the warp enable specimens to be prepared in 10 mm, 20 mm, 25 mm, 30 mm, etc, widths according to requirements

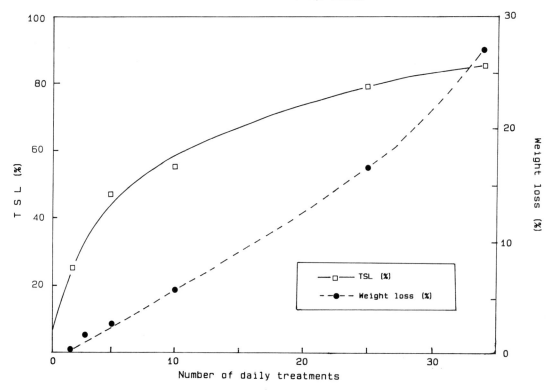

Figure 2. Loss in tensile strength (TSL) compared with loss in weight of cotton repeatedly treated with a crude cell-free cellulase preparation

adjacent white threads during preparation of the fabric. It was decided that selected vat dyes would best meet these requirements, and 2 Imperial Chemical Industry dyes in contrasting colours were finally chosen: Caledon Jade Green XN and Caledon Dark Brown BR. Yarn was package-dyed by Tootal Ltd, Sunnyside Works, Bolton, England, and the first lot of fabric was produced in August 1976. The fabric was finished in the Shirley Institute's own finishing workroom and, immediately, a problem was experienced with bleeding of the green marker threads during the caustic scour. It was quickly established that the cause lay in the reduction of the vat dyestuff to the soluble leuco form associated with the presence of reducing sugars on the cloth. The problem was solved by including Resist Salt L (sodium m-nitro benzene sulphonate) (20 g litre^{-1}) in the sodium hydroxide (20 g litre^{-1}) pad-batch scour.

When stocks of the first batch of fabric were exhausted, a further 1500 m of fabric were woven in 1981. Different dyes were used for the marker threads, and all the yarns were kier boiled prior to weaving to avoid the problem of dye bleeding. Thus, the yarn (supplied by UCO NV, Belgium) was treated by Fountain Yarn Dyers (Blackburn) Ltd, England, at 130°C for 45 minutes with sodium hydroxide (10 g litre^{-1}), Lufibrol KB (BASF West Germany) (4 g litre^{-1}), and Strodex (Dexter Chemicals, England) wetting agent (1 g litre^{-1}). The yarn was cooled to 95°C, washed twice with water at 95°C containing EDTA (5 g litre^{-1}), warm-washed with acetic acid (0.4 g litre^{-1}), followed by a cold and then a warm wash with water. The kier boiling removes any protein in the fibre lumen. Yarn for the warp marker threads was then dyed by Blackburn Yarn Dyers Ltd, England, with Solanthrene Blue RFS and Solanthrene Brown FR. (For successive cloth batches, the brown thread will be retained, but the second dye will be changed to aid recognition. The next batch will include red.)

The warp yarns were lightly sized with Courlose (sodium carboxymethyl cellulose) containing emulsified tallow and prepared on to a weaver's beam for weaving. The final fabric was scoured and carefully finished to the required width. The results for residual fat and wax content, obtained by Soxhlet extraction with methylene chloride, demonstrate the superiority of the second method of preparing the fabric, compared to that used for the earlier fabric (Table 3).

Table 3. Shirley Soil Burial Test Fabric, fat and wax content (extracted with methylene chloride)

	Extractable wet weight %
1976 fabric	0.37
1981 fabric	
Sample 1	0.21
Sample 3	0.17
Sample 5	0.20
Mean	0.19

The fabric is used to assess the activity of cellulolytic micro-organisms in the soil by tensile tests in the warp direction, using the accurately spaced marker threads woven in the fabric to define the specimen width. For example, to prepare 20 mm wide test specimens, the fabric is cut along the brown marker threads on either side of the pair of 10 mm strips defined by the green/blue marker threads. The specimen is then frayed down from the cut edges (brown markers), until one of the pair of green/blue marker ends is removed on each side; this gives a specimen nominally 20 mm wide (as defined by the green/blue markers), with a fringe of 5 mm on each side (as defined by the position of the brown cut-markers) in relation to the outer green/blue marker threads. The usual length of the individual inserted cotton strip specimens is 33 cm, and 10 replicate strips are normally used in the field. In a typical textile laboratory test, 5 specimens were tested (the minimum number allowed in British Standard 2576 (Anon 1986) is 5) to produce the results in Table 4.

Table 4. Shirley Soil Burial Test Fabric. Typical results for tensile strength (kg 20 mm^{-1} width)

Tensile strength tests (Anon 1986):
Machine – Instron, 0–100 kg range
              Gauge length, 20 cm
              Traverse speed, 5 cm min^{-1}

Specimen number	1976 fabric	1981 fabric Sample number					
		1	2	3	4	5	6
1	38.2	29.8	36.8	35.0	37.1	36.3	34.8
2	35.8	34.6	38.3	37.6	34.3	35.3	31.7
3	35.4	31.3	37.8	37.5	37.1	34.1	36.7
4	36.8	35.8	36.7	34.0	36.3	33.1	32.6
5	38.1	31.3	30.7	32.5	35.2	36.5	34.2
Mean	36.9	32.6	36.1	35.3	36.0	35.1	34.0
SD	1.28	2.52	3.07	2.22	1.23	1.45	1.95
95% CL	1.6	3.1	3.8	2.8	1.5	1.8	2.42

4 Tensile strength testing of cotton fabric

Tensile strength has long been accepted as one of the more important attributes of a woven textile. Each of the construction features of a woven cloth affects the fabric's tensile strength, but, before considering the effects of some of these features, it is important that the units and terminology are clearly defined and understood.

4.1 Terminology and textile units

In defining the following terms relating to the tensile strength testing of textiles, we have expressed a preference for SI units, but it has to be acknowledged that traditional units remain in widespread use in the textile industry.

Load: the application of a load to a specimen in its axial direction causes a tension to be developed in the specimen. The derived SI unit of force is the newton (N), being the force needed to impart an acceleration of 1 m s^{-2} to a body with a mass of 1 kg. It is an invariable quantity. The traditional way of expressing the load in gram weight (or even pound weight) is still widely used (it is common practice to leave out the word 'weight' and quote loads simply as kilograms or pounds) but, because of variations in gravitational force, a gram weight will vary in the force it may impart at different places (by about 0.5% as one moves in latitude from 0° to 90°).

Tensile strength or breaking load: the load at which the specimen breaks, expressed in newtons (N) or kg, where 1.0 kg = 9.8067 N; 1.0 N = 0.102 kg. It is essential to define the specimen width for fabric strength tests.

Stress: the ratio between force applied and the cross-sectional area of the specimen. The preferred SI units are kN m^{-2}.

Mass stress: because the cross-sections of many fibres and fabric structures are irregular in shape and difficult to measure, it is preferable to use the linear density of the specimen (a dimension related to cross-section), expressed in tex, to establish the mass stress unit, cN tex^{-1} (centinewtons per tex).

Tenacity: the mass stress at break (cN tex^{-1}). Expressing the breaking strengths of different materials in terms of tenacity permits direct comparisons between materials of varying fineness. To convert the old units of g tex^{-1} to cN tex^{-1}, multiply g tex^{-1} by 0.981.

Strain: application of a load to a specimen causes it to stretch by an amount which varies with the initial length of the specimen; the strain is the ratio of the elongation to the initial length. The shape of the stress versus strain curve for a textile fibre is largely governed by its molecular structure. Typical shapes of stress/strain curves for various textile fibres are shown in Figure 3.

Extension: the elongation expressed as a percentage of the initial length.

Breaking extension: the extension of the specimen at breaking point.

Initial or Young's modulus: because the overall stress/strain curves of textile materials are not linear, the breaking stress gives no indication of the mechanical behaviour at low loads (low stress) and small extensions (low strains), which govern how a textile structure feels, drapes, and deforms. The appropriate measure for such small deformations is the initial or Young's modulus, given by the linear relationship between stress and strain in the first portion of the stress/strain curve where the material behaves elastically. The SI unit for initial modulus is the same as for breaking stress, ie N tex^{-1}, but it must be remembered that it is essentially a theoretical value produced by projecting the initial slope of the stress/strain curve to an imaginary strain of 1.0, and gives much higher numerical values than the actual breaking stress.

Work of rupture: the energy or work required to break the specimen, given by the area under the stress/strain curve. The work of rupture value is an indication of the resistance of the material to sudden shock.

4.2 Constructional features affecting the tensile properties of fabrics

Whilst each of the constructional features of a woven fabric affects tensile strength, some of these features are important and obvious, whereas the effects of others are negligible or obscure. Certainly, the relationship of the strength of a fabric to its construction is too involved and insufficiently understood in detail to be expressed in a single formula. The following attempts to indicate briefly some of the more important relationships between construction and strength.

4.2.1 *Fibre quality and fabric weight*

Two most important factors are clearly the quality of the fibre and the total quantity of cotton fibre in the direction under test. The integral strength of the fibres in a cross-section of the cloth gives an upper limit to the strength that could possibly be achieved, but the realized strength of a fabric is usually only about 50% of this figure, because of fibre slippage and irregularity of fibre distribution along the yarn. The integral strength of the yarns is a more useful guide than fibre strength, and experience shows that the tensile strength of fabrics of common construction is usually between 85% and 125% of the integral strength of all the yarns in the direction tested.

4.2.2 *Yarn strength*

In order to examine the relationship between fabric tensile strength and integral yarn strength, imagine a test specimen from which all the crossing threads have been removed, the longitudinal threads being gripped as a band between parallel jaws. The threads fail one by one in order of increasing extensibility as the jaws are moved apart. The coefficient of variation of single-thread breaking extension of cotton yarns is

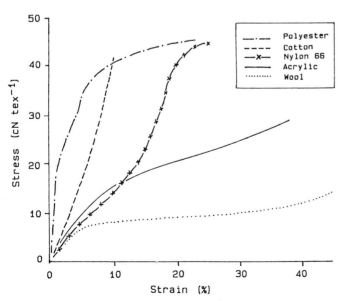

Figure 3. Typical shapes of stress/strain curves for various textile fibres

typically about 12%. With this amount of variation in breaking extension, a band of threads might be expected to have a strength of only about 75% of the integral strength of the threads but, in practice, the strength of a fabric strip is nearly always considerably greater because of the presence of the crossing threads interlaced with the threads under tension.

4.2.3 *The effects of the crossing threads*
Taylor (1959) discusses in detail 3 effects associated with the presence of the crossing threads. Briefly, their presence (i) localizes the rupture, (ii) causes crimping (the bending imposed on yarns by their interlacing in a woven structure) of the longitudinal threads so that they do not lie continuously in the direction of stress, and (iii) increases the binding of the fibres in the individual yarns, thereby increasing yarn strength. The first effect can be likened to the band of threads being gripped at short intervals along the length, which reduces the effective test length, thereby raising the observed fabric strength by an amount depending on the tightness of the weave. Yarn in a woven fabric must be crimped, and the presence of crimp in the longitudinal threads has been found to decrease the cloth strength ratio (the ratio of fabric strength to the average single-thread strength multiplied by the number of longitudinal threads in the specimen) by an average of 1% for each 1% increase in crimp.

4.2.4 *Yarn twist*
Fabric strength is related to the twist in the yarn because of its influence on yarn strength. The strength of a spun cotton yarn is, of course, less than the integral strength of the component fibres in the cross-section of the yarn because of the obliquity of the fibres to the yarn axis, fibre slippage, and the fact that all the fibres do not break simultaneously.

4.2.5 *Threads per metre and yarn count*
Any increase in the density of a weave will generally tend to improve fabric tensile strength relative to yarn strength, associated with improved fibre binding, but, if the yarns are already very well bound, as in the case with folded or highly twisted yarns, the effects may be small or offset by the accompanying increased yarn crimp.

4.2.6 *Effect of weave*
The strongest weaves are generally found to be those with the greatest number of intersections in the weave repeat, ie plain and 2/1 twill. However, the magnitude of this effect depends on the type of cotton and the yarn construction; with folded yarns twisted for optimum yarn strength, the deleterious effect of the greater yarn crimp associated with the plain weave might overwhelm any advantages to be gained from fibre binding.

5 *Testing features affecting the results obtained*
5.1 The effects of humidity and temperature
The influence of moisture on the mechanical properties of textile materials depends on the type of fibre. Whereas hydrophobic material such as polypropylene or polyester may be little affected, the hydrophilic cellulose fibres, including cotton, exhibit significant differences in their stress/strain properties when tested under dry and wet conditions. A hysteresis effect on tensile strength is shown during wetting/drying cycles. It is essential to carry out routine testing in a laboratory with a standard atmosphere of 65 ± 2% relative humidity and 20 ± 2°C. Sufficient time (24 h) should be allowed for the sample to reach equilibrium with the standard atmosphere before the tests are carried out, and fabric samples for breaking strength should be preconditioned at a relative humidity not exceeding 10% at a temperature of 50°C for 4 h and then conditioned in a standard atmosphere for 24 h. (Some users of the cotton strip assay, however, do tensile testing on wet specimens to obviate humidity control (Wynn-Williams 1988).)

5.2 Test specimen length
The breaking load (S_1) recorded for a specimen corresponds to that of the weakest cross-section along the specimen's length, ie the position where the aggregate strength of the longitudinal threads in the specimen width is weakest. If the specimen had been tested in 2 halves, 2 breaking loads would have been obtained, S_1 and S_2, the mean of which would have been higher than S_1. Hence, because of this 'weak link' effect, the apparent tensile strength can be increased by testing the material at a shorter gauge length (ie initial jaw separation).

5.3 Rate of loading and time to break
The more rapid the rate of loading, the higher the breaking load. Midgley and Pierce (1926) established an empirical relationship between the tensile strength values obtained and the time taken to break the specimens (fibres, yarns, and fabrics). The increase in strength amounts to about 10% for a 10-fold decrease in time taken to break the specimen.

6 *The mechanics of tensile stength testing machines*
There are 3 basic methods of loading the specimen to observe the effects of tensile forces.

6.1 Constant rate of traverse (CRT)
When the load is applied on a pendulum lever machine, both jaws move as the load is applied. The rate of loading and the time to break are influenced by the extension of the specimen. The velocity of the upper jaw is less than that of the lower jaw by an amount dependent upon the extensibility of the material.

6.2 Constant rate of loading (CRL)
The specimen is gripped in a fixed top jaw and in a bottom jaw which is movable. A force is applied at a constant rate, extending the specimen until it eventually breaks. Clearly, the loading causes the elongation.

6.3   Constant rate of extension (CRE)
In this type of instrument, the specimen is again gripped in a fixed top jaw and in a bottom jaw which, in this case, can be moved downwards at a constant velocity by means of a screw mechanism. As the bottom jaw moves downwards, the specimen is extended and tension is increased until the specimen finally breaks. Clearly, the cause and effect are now the other way round compared to the CRL instruments; with the CRE instrument, it is the extension which causes the loading. The same principles obtain with those instruments where the specimen is mounted horizontally.

Figure 4 illustrates that, when tested under CRE conditions, the tension developed in the specimen is almost 70% of the breaking strength after only 30% of the time to break has elapsed, and the specimen therefore spends most of the test at the higher loads. In contrast, under CRL conditions, the specimen spends only 30% of the time to break under a load which is above 70% of the breaking load.

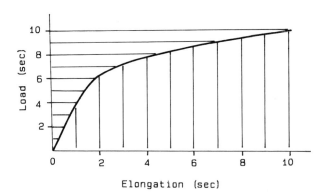

Figure 4.   Comparison of constant rate of extension (elongation) and constant rate of loading (load) testing conditions

6   *References*

Anon. 1981. *Method of test for the determination of the resistance of textiles to microbiological deterioration.* (BS 6085.) London: British Standards Institution.
Anon. 1986. *Methods of test for textiles - woven fabrics - determination of breaking strength and elongation (strip method).* (BS 2576.) London: British Standards Institution.

Kassenbeck, P. 1970. Bilateral structure of cotton fibres as revealed by enzymatic degradation. *Text. Res. J.,* **40,** 330-334.
Midgley, E. & Pierce, F.T. 1926. Tensile tests for cotton yarns – the rate of loading. *J. Text. Inst.,* **17,** 330-341.
Taylor, H.M. 1959. Tensile and tearing strength of cotton cloths. *J. Text. Inst.,* **50,** 161-188.
Wynn-Williams, D.D. 1988. Cotton strip decomposition in relation to environmental factors in the maritime Antarctic. In: *Cotton strip assay: an index of decomposition in soils,* edited by A.F. Harrison, P.M. Latter & D.W.H. Walton, 126-133. (ITE symposium no. 24.) Grange-over-Sands: Institute of Terrestrial Ecology.

Microbial cellulases and their action on cotton fibres

B F SAGAR
Shirley Institute, Manchester

1 Summary
Micro-organisms producing cellulolytic enzymes are found amongst the bacteria, the actinomycetes and, particularly, the microfungi. Whereas the ability to break down cellulose is very widespread amongst the latter group of micro-organisms, relatively few microfungi elaborate extracellular enzymes that can degrade highly ordered 'crystalline' cellulose.

It is now well established that the cellulase system, which attacks highly ordered native cellulose, is a group of enzymes acting in concert, but the actual mechanism of the synergistic action between the so-called C_1 and C_x components in the cellulase complex remains uncertain. Exhaustive attempts to detect changes in the fine structure of cotton cellulose, brought about by the individual and combined action of the C_1 and C_x components of the cellulase complexes from various microfungi, have confirmed that this highly ordered substrate is unaffected, unless both major enzyme components are present. The combined action of C_1 and C_x appears to be confined to the pair of surfaces of the elementary fibrils in the cotton fibre, which contain cellulose molecules with a specific spatial disposition of the 2, 6- and 2, 3, 6- hydroxyl groups accessible on alternate anhydroglucose units along the chain.

2 Introduction
Many micro-organisms produce extracellular enzymes which catalyse the hydrolysis of water-soluble cellulose derivatives (eg carboxymethyl- and hydroxyethyl-cellulose of relatively low degrees of substitution) and of highly swollen forms of cellulose (eg phosphoric acid-swollen cellulose), but relatively few produce cell-free enzyme preparations that have the ability to hydrolyse highly ordered crystalline cellulose, as typified by cotton. Notable in this regard are the cellulases of *Trichoderma viride* (Selby & Maitland 1967), *T. reesei* (Mandels & Reese 1964), *T. koningii* (Wood 1968), *Penicillium funiculosum* (Selby 1968), *Fusarium solani* (Wood 1969), and the basidiomycete *Sporotrichum pulverulentum* (Eriksson 1978). Most native lignocellulosic materials are difficult to break down enzymically; enzymolysis is impeded by the presence of the lignin component and by the morphological fine structure of these materials. Cotton is free from the difficulties associated with lignin. Nevertheless, the rate of enzymolysis of scoured cotton is still relatively slow because of its highly ordered, hydrogen-bonded, fine structure. Once the cotton fibre has dried in the opened boll, it becomes a most searching substrate for evaluating the efficiency of any cellulase system and for investigating the mechanism of the cellulase action. Inaccessibility of native cotton to the large

protein molecules plays a major role in regulating the cellulase action. Compared with hydrolysis by mineral acids, greater losses of weight are observed during enzymic attack relative to the production of a given number of reducing chain ends in the insoluble residue. Enzymic breakdown spreads from relatively fewer points of initial attack.

3 The cellulase complex
Fractionation studies carried out in a number of different laboratories on cell-free enzyme preparations, obtained from highly active cellulolytic microfungi mentioned above, have shown the presence of 3 different basic types of enzyme in every case (Table 1). Considerable effort has been devoted over the past decade to the isolation, purification, and characterization of the C_1, C_x, and β-D-glucosidase enzymes, resulting in controversy over the role of C_1. These 3 components of the cellulase complex achieve the breakdown of cotton when they act in concert; they lose this ability when separated, but recover it again when recombined in their original proportions (Table 2). There are now several examples of the C_1 com-

Table 1. The components of the cellulase complex

1.	C_1:	Action still not fully understood. Deaggregates cellulose molecules on the surface of the elementary fibrils in cotton
2.	C_x:	β-1, 4-glucanases. The 'x' reflects the multiplicity of these components
		i. Exo-enzyme; splits off (a) glucose (b) cellobiose from the non-reducing chain ends
		ii. Endo-enzyme; acting randomly along the chain. Terminal linkages more resistant. Hydrolyzes water-soluble cellulose derivatives. No action on cotton in absence of C_1
3.	β-glucosidases	Hydrolyze cellobiose and short-chain oligosaccharides to glucose

Table 2. Synergistic action resulting from recombination of cellulase fractions

Component	Relative cellulase activity (%)
Original solution	100
C_1	1
C_x	5
$C_1 + C_x$	102
CMC-ase	44
Cellobiase	<1
CMC-ase + cellobiase	2
C_1 + CMC-ase	35
C_1 + cellobiase	20
C_1 + CMC-ase + cellobiase	104

ponent from one microfungus acting synergistically with the C_x of a different microfungus to solubilize native cotton. Selby (1968) was the first to report this particular kind of 'crossed synergism' with the C_1 component of *P. funiculosum* and the C_x of *T. viride* (Figure 1), and Wood and McCrae (1979) have reported similar effects with different combinations of the C_1 and C_x components from *T. koningii*, *F. solani* and *P. funiculosum*.

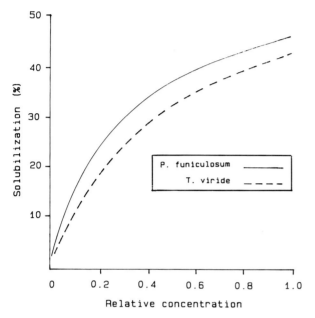

Figure 1. *Crossed synergism involving the C_1 component of* Penicillium funiculosum *and the C_x components of* P. funiculosum *and of* Trichoderma viride

4 The importance of the fine structure of cotton fibres

It is now abundantly evident that the important factor determining the susceptibility of cotton cellulose fibres to enzymic attack is the accessibility of the surfaces of the microfibrils that make up the primary and secondary cell wall material. Degradation starts in the more accessible parts of the fibre, identified by Kassenbeck (1970), and these regions tend to be completely removed before degradation of the zones with a higher density of fibrillar packing takes place. However, we have to reconcile these observations with the fact that even the more accessible microfibrils must be considered to be almost entirely crystalline, and any attack by enzymes is presumed to occur by surface erosion of the individual microfibrils. The rate of solubilization of cotton by cell-free cellulase preparations from *T. viride* and *P. funiculosum* follows first-order kinetics with respect to substrate concentration (Selby & Maitland 1967), indicating that the reactivity of the microfibrillar surfaces is uniform and the available surface area is proportional to the amount of substrate remaining throughout the degradative process. The fact that there is no sharp change in the slope of the first-order plots indicates that the fresh surfaces exposed as the attack proceeds have the same reactivity (Figure 2).

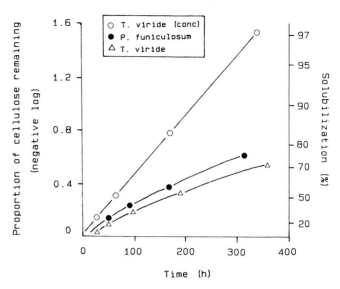

Figure 2. *Kinetics of solubilization of cotton by cellulases from* P. funiculosum *and* T. viride

There are 3 hydroxyl groups, at C-2, C-3 and C-6, on each anhydroglucose unit (agu) along the cellulose chains that are potentially available for binding with the cellulase enzymes but, because of the aggregation of the cellulose molecules into close-packed, hydrogen-bonded structural units, not all of these hydroxyl groups are available on all the surfaces of the microfibrils. From studies of repeated methylation of cotton with dimethyl sulphate, followed by hydrolysis of the partially methylated cellulose and gas liquid chromatographic analysis of the trimethylsilyl derivatives of the derived partially methylated glucoses, we deduced (Haworth *et al.* 1969) a model for the elementary structural unit in cotton cellulose. It contained, on average, a bundle of 80 cellulose chains in a block whose cross-section contained 8 x 10 agu units, with one pair of opposite surfaces containing only the 2- and 6- hydroxyl groups accessible (on alternate agu) and with the other pair of opposite surfaces with the 2, 6- and the 2, 3, 6- hydroxyl groups available. The presence of larger units (averaging 16 x 10 agu) was indicated by a similar examination of the methoxyl distribution in cotton methylated with diazomethane in ether saturated with water, suggesting that these elementary fibrils must be aggregated under such conditions. Indeed, in the dry state, cotton must be almost completely aggregated, as dry cotton is scarcely methylated at all by dry ethereal diazomethane. Water must, therefore, lead to extensive deaggregation, but complete disruption to the ultimate elementary fibrils probably only occurs in the presence of a strong swelling agent, such as 2N NaOH. However, by causing partial deaggregation of the fibrils, wetting cotton fabric strips must result in an increase in the surface area available for enzyme attack.

Enzymic attack is believed to occur only on the surfaces in which the 2, 6- and 2, 3, 6- hydroxyl groups are accessible (Figure 3).

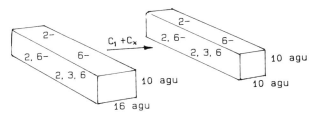

Figure 3. The combined action by C_1 + C_x appears to be centred on the surfaces of the elementary fibrils in cotton which contain the 2, 6- and 2, 3, 6- hydroxyl groups

5 The role of C_1

Whereas the C_x components of the cellulase complex exhibit the classical exo- and endo-β-1, 4-glucanase activities (Wood 1985), the action of the C_1 component remains unclear. In the original concept of Reese *et al.* (1950), C_1 was believed to act first on the highly ordered arrays of cellulose molecules in the elementary fibrils, by deaggregating the cellulose chains (C_1 has been described as a hydrogen-bondase), and making them available for subsequent hydrolysis by the β-1, 4-glucanases. Following the development of improved fractionation methods and the discovery that the C_1 fraction appears to possess cellobiohydrolase activity, several groups of workers reversed this original concept by suggesting that it is the endo-β-1, 4-glucanase which attacks first, in a random manner, to produce chain ends for subsequent attack by the exo-enzyme, β-1, 4-glucan cellobiohydrolase. However, Reese (1976) critically questioned the implication that C_1 and this cellobiohydrolase are one and the same enzyme activity, and gave pertinent reasons for rejecting this suggestion. In the light of the new evidence, he did, however, modify the original concept to include the possibility that C_1, acting as an endo-enzyme, randomly splits the glucosidic bonds of cellulose molecules situated on the surface of the elementary fibrils.

This expanded concept makes C_1 a member of the C_x random-acting enzymes, albeit a special member with properties not possessed by any previously defined C_x component. Its action is confined to 'crystalline' cellulose and, unlike the endo-C_x enzyme (β-1, 4-glucan glucanohydrolase), it has no action on cellulose derivatives.

The fact that the requirement for the C_1 enzyme is considerably reduced when highly ordered cellulose is rendered more accessible by physical means, such as ball-milling or swelling, is at variance with the idea that the only activity possessed by the C_1 enzyme is that of a β-1, 4-glucan cellobiohydrolase. It is axiomatic that, if a cell-free cellulase preparation acts on native cotton, then it must contain C_1.

We have used several techniques (Sagar 1978) in an attempt to detect any changes in the fine structure of native cotton after treatment with the individual C_1 or

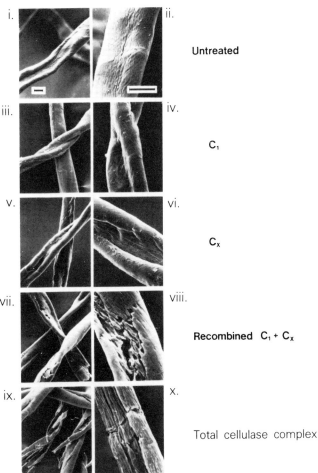

Plate 2. Scanning electron micrograph of fibres treated with C_1 and C_x enzymes separately and in combination

C_x components of the cellulase complex obtained by conventional chromatographic separation, but without success, confirming that this highly ordered cellulose is totally unaffected, unless both enzymes are present; see, for example, the appearance of variously treated fibres under a scanning electron microscope (Plate 2). The observation that the action of the cellulase complex appears to be confined only to the pair of faces of the microfibrils containing the 2, 6- and 2, 3, 6- hydroxyl groups led to the suggestion (Sagar 1978, 1985) that the C_1 component has a specific affinity for the cellulose molecules in this pair of surfaces (associated with the unique spatial configuration of their accessible hydroxyl groups), leading to chain scission and production of free cellulose chain ends accessible to the C_x exo-glucanases. The proposed actions of the various components of the cellulase complex are summarized in Figure 4.

The possibility must be considered that this unique stereospecific endo-glucanase activity (which we shall continue to call C_1) resides in the same protein molecule as that containing the cellobiohydrolase activity. Alternatively, it is possible that the C_1 activity is mediated by an enzyme–enzyme complex formed between the β-1, 4-glucan cellobiohydrolase and one of

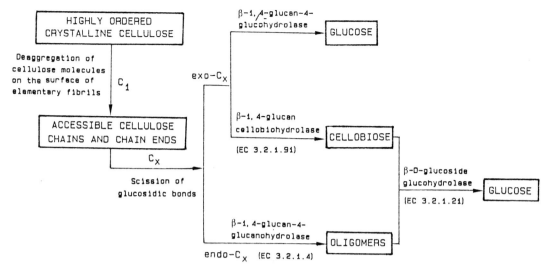

Figure 4. The role of each component of the cellulase complex. EC numbers of enzyme nomenclature given in brackets

the other C_x enzymes (Sagar 1978, 1985; Wood & McCrae 1978). Significant advances continue to be made in understanding the mechanism of cellulase enzymolysis, but several important questions remain unanswered, relating to the stereospecific arrangement of the cellulose molecules on the microfibrillar surfaces and the interaction of the endo- and exo-glucanases.

6   References

Eriksson, K.E. 1978. Enzyme mechanisms involved in cellulase hydrolysis by the rot fungus *Sporotrichum pulverulentum. Biotechnol. Bioeng.,* **20,** 317-332.

Haworth, S., Jones, D.M., Roberts, J.G. & Sagar, B.F. 1969. Quantitative determination of mixtures of alkyl ethers of D-glucose Part II. Structural studies of partially methylated cotton cellulose. *Carbohyd. Res.,* **10,** 1-12.

Kassenbeck, P. 1970. Bilateral structure of cotton fibres as revealed by enzymatic degradation. *Text. Res. J.,* **40,** 330-334.

Mandels, M. & Reese, E.T. 1964. Fungal cellulases and the microbial decomposition of cellulosic fabric. *Develop. ind. Microbiol.,* **5,** 5-20.

Reese, E.T. 1976. History of the cellulase programme at the U.S. Army Natick Development Center. *Biotechnol. Bioeng. Symp.,* no. 6, 9-24.

Reese, E.T., Siu, R.G.H. & Levinson, H.S. 1950. The biological degradation of soluble cellulose derivatives and its relationship to the mechanism of cellulose hydrolysis. *J. Bacteriol.,* **59,** 485-497.

Sagar, B.F. 1978. *Investigation of accessibility changes in 'crystalline' cellulose.* (AD-A058080.) Springfield: National Technical Information Service.

Sagar, B.F. 1985. The mechanism of cellulase action. In: *Cellulose and its derivatives: chemistry, biochemistry and applications,* edited by J.F. Kennedy, G.O. Phillips, D.J. Wedlock & P.J. Williams, 199-207. Chichester: E. Horwood.

Selby, K. 1968. Mechanism of biodegradation of cellulose. In: *Biodeterioration of materials, microbiological and other aspects,* edited by A.H. Walters & J.J. Elphick, 62-78. Amsterdam: Elsevier.

Selby, K. & Maitland, C.C. 1967. The cellulase of *Trichoderma viride.* Separation of the components involved in the solubilization of cotton. *Biochem. J.,* **104,** 716-724.

Wood, T.M. 1968. Cellulolytic enzyme system of *Trichoderma koningii. Biochem. J.,* **109,** 217-227.

Wood, T.M. 1969. Cellulase of *Fusarium solani.* Resolution of the enzyme complex. *Biochem. J.,* **115,** 457-464.

Wood, T.M. 1985. Aspects of the biochemistry of cellulose degradation. In: *Cellulose and its derivatives: chemistry, biochemistry and applications,* edited by J.F. Kennedy, G.O. Phillips, D.J. Wedlock & P.J. Williams, 173-188. Chichester: E. Horwood.

Wood, T.M. & McCrae, S.I. 1978. The cellulase of *Trichoderma koningii. Biochem. J.,* **171,** 61-72.

Wood, T.M. & McCrae, S.I. 1979. Synergism between enzymes involved in the solubilization of native cellulose. In: *Hydrolysis of cellulose: mechanisms of enzymatic and acid catalysis,* edited by R.D. Brown & L. Jurasek, 181-209. Washington, DC: American Chemical Society.

Standardization of rotting rates by a linearizing transformation

M O HILL[1], P M LATTER[2] & G BANCROFT[3]
[1]Institute of Terrestrial Ecology, Monks Wood Experimental Station, Huntingdon
[2]Institute of Terrestrial Ecology, Merlewood Research Station, Grange-over-Sands
[3]Aynsome Laboratories, Grange-over-Sands

1 Summary

Linearization is a mathematical technique for inferring process rates from an observed response variable that, under constant conditions, does not change linearly with time. Experiments have shown that, under constant conditions, tensile strength of buried cotton cloth changes according to the relation:

$$y = y_0/ (1 + (CRR.t)^3)$$

where y_0 and y are initial and final tensile strength, t is time, and CRR is the cotton rotting rate. In soil insertion tests, the initial and final tensile strengths (TS) of cotton are known, so the loss (CTSL) can be calculated, and CRR yr^{-1} may be estimated from the formula:

$$CRR = \sqrt[3]{(CTSL/\text{final TS})} \times 365 / t$$

where t is the duration of insertion in days. Thus, using CRR, degradation rates can be manipulated freely, eg to derive a mean annual value, time to 50% CTSL (also used for estimating retrieval time), or a temperature response coefficient Q_{10}.

2 Introduction

When a cotton strip is inserted in a particular soil, the reason commonly given is that the research worker aims to determine the potential for cellulose degradation under particular environmental conditions.

This vague answer requires elucidation. By 'potential' is meant the potential rate, which assumes that there is such a thing as a general rate of cellulose degradation. In one sense, a generalized rate is a meaningless hypothetical construct; much depends on how cellulose is presented to decomposer organisms in the soil. However, if, to a reasonable approximation, the rate for one type of substrate is a multiple of that for another, then results for the rate of degradation of a cotton strip could be generalized to materials such as leaves and rotten wood.

Underlying this idea is a multiplicative model (cf Swift *et al.* 1979, p259). In symbols:

$$R(T, M, Q,...) = \text{const} \times f(T) \times f(M) \times f(Q) \times ...$$

where R is the rate of the decomposition process, and T, M, Q, .. are variables such as temperature, moisture and substrate quality, which determine the value of R. The advantage of the cotton strip method for soil assay is that it fixes the value of $f(Q)$, allowing the effects of the other variables to be determined more accurately.

3 Need for linearization

Unfortunately, the cotton strip assay does not lend itself naturally to the definition of a process rate, R, in contrast, for example, with respirometry, for which the rate of oxygen uptake defines a natural measure of the rate at which the process is occurring. The purpose of linearization is to convert an arbitrary response variable, which might be tensile strength loss, mass loss, FDA hydrolysis (Smith & Maw 1988), or some such factor, to a derived variable that changes linearly with time.

It is instructive to consider an analogous problem familiar to ecologists, namely how to define a process rate for the decay of organic matter in litter bags. Suppose that 100 g of litter are placed in a bag and that, after one year, 50 g remain. This phenomenon would very likely be described by saying that the decay rate:

$$k = \log_e (100/50) = 0.69 \text{ g yr}^{-1}$$

Underlying this description is a model of decay under constant conditions, namely that the proportional rate of loss is constant.

In symbols:

$$y = y_0 \exp(-kt)$$

where y is the measured response variable (mass remaining in the bag), and k is a constant for those environmental conditions, called the decay rate.

Now, it is well known that, under field conditions, the actual instantaneous rate of decay will vary in response to temperature, moisture, and other environmental influences. The decay rate, k, is thus not in reality a constant, but an estimate of the average rate of decay over the year. This average value can be treated as a constant feature of the site, because the between-year variation will usually be small compared to seasonal variation within a year.

To estimate k, we take logarithms:

$$k = \log_e (y_0/y)/t$$

Note that $\log_e$ is the inverse function of exp, ie if $y = \exp(x)$, then $x = \log_e(y)$. Thus, the process rate, k, is estimated by using the inverse function ($\log_e$) of the function (exp) which defines the change in y under constant conditions.

In order to perform a linearizing transformation of this kind, one essential condition must apply, namely that, under constant conditions,

$$y/y_0 = f(Rt)$$

where R is a rate parameter and f is some function. In other words, the shape of the curves describing variation in y over time must be independent of the experimental conditions, although the rate at which things happen (ie the parameter R) may vary. If the shape of the curve $f(x)$ is known, then it is possible to estimate the rate parameter from the equation

$$R = f^{-1}(y/y_0) / t$$

where f^{-1} denotes the inverse function of f.

4 Hueck–Toorn degradation curves

Hueck and Toorn (1965) made a study of the form of the decay curve for loss of cotton tensile strength under constant conditions in soil burial beds at 28°C. They fitted curves of the form

$$y = y_0 / (1 + (t/t_{50})^b)$$

to results of 11 individual experiments. Their model has 3 parameters: y_0, the initial tensile strength; t_{50}, the time to 50% loss of tensile strength; and b, a parameter specifying the shape of the curves. For untreated cloth, they found mean parameter values

$$y_0 = 54 \text{ kg cm}^{-1} \text{ (53 kN m}^{-1}\text{)}$$
$$t_{50} = 3.1 \text{ days}$$
$$b = 3.0$$

It can be seen that the shape of the curves is not very sensitive to variations in the parameter b (Figure 1). However, if b varies, then so, by implication, does the shape of the decay curves, and linearization is not possible.

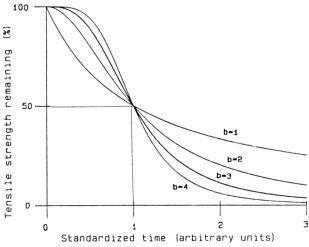

Figure 1. Curves of the Hueck–Toorn family, standardized to 50% loss of tensile strength at time $t = 1$. The curves have the formula $TS = 100/(1 + t_b)$. The parameter b determines the shape of the curves

5 An experiment to compare soils

Following Hueck and Toorn's work, cotton strips were laid within trays of contrasting soils out of doors at the Institute of Terrestrial Ecology's Merlewood Research Station, in Cumbria. Detailed results have been published elsewhere (Hill et al. 1985). The aim was to answer 2 questions: did the curves defining the rate of loss of tensile strength have approximately the same shape, and, if the shape was approximately similar, how could it be parameterized?

The experiment was not conducted under controlled conditions. For cotton buried in raised-bog peat, it was necessary to wait more than 2 years before the degradation process was complete. It was, thus, necessary to make allowance for a reduced rate of rotting during the winter, one-third of the summer rate, and the results for 5 soils (Figure 2) then agreed well with those of Hueck and Toorn (1965). Furthermore, it was possible to confirm that the value $b = 3.0$ fitted our data as well as those of Hueck and Toorn.

On this basis, it is possible to describe the change in tensile strength over time by an equation of the right functional form for linearization, namely:

$$y/y_0 = \text{proportion of tensile strength remaining}$$
$$= (1 + (CRR.t)^3)^{-1}$$

where CRR is a single parameter defining the process rate, and is, by definition, $CRR = CT50^{-1}$, with CT50 = time to 50% CTSL.

Values of CRR range from 1.0 yr^{-1} (CT50 = 365 days) to 40 yr^{-1} (CT50 = 9.1 days) (Ineson et al. 1988), with antarctic peat soils giving the lowest values and a tropical swamp the highest.

6 The linearizing transformation

The behaviour of tensile strength over time has now been described by a parametric relation of the form:

$$y/y_0 = f(CRR.t)$$

where $f(x) = 1/(1 + x^3)$.

To estimate the process rate, CRR, we need to know the inverse function of f. This is given by:

$$f^{-1}(y) = \sqrt[3]{(1 - y)/y}$$

Let CR (cotton rottenness) be defined by:

$$CR = f^{-1}(y/y_0)$$
$$= \sqrt[3]{(y_0-y)/y}$$
$$= \sqrt[3]{(CTSL/final\,TS)}$$

where CTSL = initial (or field control) TS – final TS. Then, the process rate $CRR = CR/t$.

Provided that the extreme values (unrotted or totally

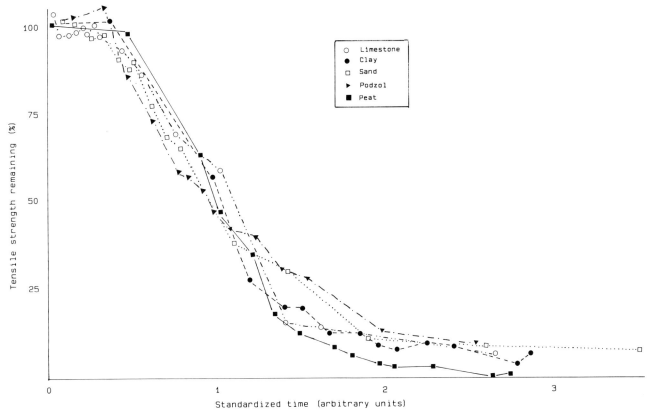

Figure 2. *Loss of tensile strength in relation to time in trays of soil at ITE's Merlewood Research Station. The curve for each type of soil has been standardized by adjusting the timescale, so that, as far as possible, the curves lie on top of one another. Allowance has also been made for a reduced rate of rotting in winter (cf Hill et al. 1985)*

rotted) of y/y_0 were avoided, independent estimates of CRR were consistent for each soil, with a coefficient of variation of 11%. However, for values of y/y_0 outside the range 0.1–0.9, estimates of CRR were much less reliable, as also discussed by Walton (1988). As a test of the linearizing transformation, CR was estimated from CTSL after differing periods of burial (Figure 3).

It is now recommended that CRR should be expressed in annual units, even if CT50 is as low as 3 days, which is the sort of value obtained from soil burial beds at 30°C. If CT50 = 3 days, then CRR = 365/CT50 = 122 yr^{-1}. This value may be compared directly with the slower rate of 21 yr^{-1} obtained out of doors at 14°C, suggesting that Q_{10} for the process rate is about 3.0.

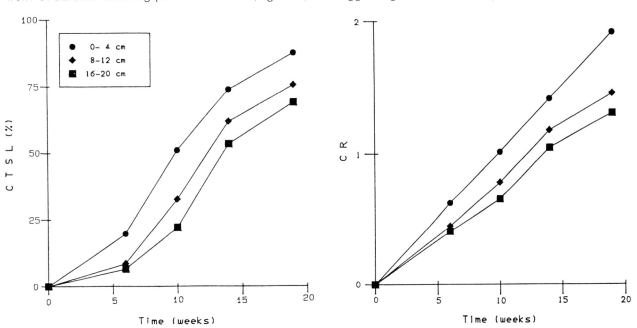

Figure 3. *Loss in tensile strength in relation to time in soil in a field experiment at Gisburn Forest, Lancashire (Brown & Howson 1988). Combined data for 4 monoculture plots are presented as unlinearized (CTSL %) and linearized (CR) according to formula in text*

7 Applications

Given the linearizing transformation, attention can be focused on the results of cotton strip assays, and their interpretation. In particular, it is possible to estimate an annual rate of rotting for a site, based on a number of individual observations. Suppose that strips are buried at 4-monthly intervals, say on 1 January, 1 May and 1 September, and recovered after 6 weeks. If the seasonal rates of rotting, CRR, are 3 yr^{-1} for January, 5 yr^{-1} May and 8 yr^{-1} September, then an estimate of the mean annual value is $(3 + 5 + 8)/3 = 5.3$ yr^{-1}. The meaning of this value is that, if each strip were left in place until it had reached 50% CTSL, and then withdrawn and replaced by a fresh one, 53 strips would be decomposed in 10 years. This annual value allows a direct comparison with arctic sites, where it is possible to leave a single set of strips buried for a whole year before recovery.

A mean result can also be calculated for a sequential sampling series, where strips are inserted at one time but removed at several time intervals (provided that the means for any one removal date are within the range 10–90% CTSL).

The optimum retrieval time of approximately 50% loss is easily estimated using a CT50 calculated for a set of test control samples at a certain time t, where:

$$\text{days to 50\% CTSL} = \frac{\text{days at time t}}{\text{CR at time t}}$$

Using the linearized process rate, CRR, it is also possible to define a temperature response Q_{10} for decomposition. If observations are available from tests at differing temperatures, and if the soils are not too different, then the decomposition rate may be expected to show a roughly exponential temperature response:

$$\text{CRR} = \text{constant} \times Q_{10}^{(T/10)}$$

ie $\log_e \text{CRR} = \text{constant} + \log_e \text{CRR} = \text{constant} + \log_e (Q_{10})/10 \times T$.

In other words, if the slope of the regression of $\log_e$ CRR on temperature is b, then:

$$Q_{10} = \exp(10 \times b)$$

8 References

Brown, A.H.F. & Howson, G. 1988. Comparison of cellulose decomposition in stands of 4 tree species using cotton strips. *Biol. Fertil. Soils.* In press.

Hill, M.O., Latter, P.M. & Bancroft, G. 1985. A standard curve for inter-site comparison of cellulose degradation using the cotton strip method. *Can. J. Soil Sci.,* **65,** 609-619.

Hueck, H.J. & Toorn, J. van der. 1965. An inter-laboratory experiment with the soil burial test. *Int. Biodeterior. Bull.,* **1,** 31-40.

Ineson, P., Bacon, P.J. & Lindley, D.K. 1988. Decomposition of cotton strips in soil: analysis of the world data set. In: *Cotton strip assay: an index of decomposition in soils,* edited by A.F. Harrison, P.M. Latter & D.W.H. Walton, 155-165. (ITE symposium no. 24.) Grange-over-Sands: Institute of Terrestrial Ecology.

Smith, R.N. & Maw, J.M. 1988. Relationships between tensile strength and increase in metabolic activity on cotton strips. In: *Cotton strip assay: an index of decomposition in soils,* edited by A.F. Harrison, P.M.Latter & D.W.H. Walton, 55-59. (ITE symposium no. 24.) Grange-over-Sands: Institute of Terrestrial Ecology.

Swift, M.J., Heal, O.W. & Anderson, J.M. 1979. *Decomposition in terrestrial ecosystems.* Oxford: Blackwell Scientific.

Walton, D.W.H. 1988. The presentation of cotton strip assay results. In: *Cotton strip assay: an index of decomposition in soils,* edited by A.F. Harrison, P.M. Latter & D.W.H. Walton, 28-31. (ITE symposium no. 24.) Grange-over-Sands: Institute of Terrestrial Ecology.

Some statistical problems in analysing cotton strip assay data

D K LINDLEY and D M HOWARD
Institute of Terrestrial Ecology, Merlewood Research Station, Grange-over-Sands

1 Summary

The statistical methods suitable for data derived from cotton strip assay are those which are also appropriate for soil profile data, and the problems encountered are similar. Precision can be increased by using stratified sampling for survey work and blocking for designed experimental studies. The need for adequate replication of the experimental plots, rather than increasing the number of replicate strips, is emphasized. Repeated measurement techniques will be required to analyse depth and time effects, and multiple regression may help to explain differences between sites.

2 Introduction

Some of the statistical problems in analysing data from the cotton strip assay are inherent, whilst some are enlarged by the observer or experimenter. The distinction between these 2 types is made because the observer obtains the information by a survey approach, whereas the experimenter collects the information from a designed experiment.

The problems can be grouped under the following headings:

  i. within site variability;
  ii. problems with depth and time;
  iii. the use of appropriate experimental designs;
  iv. use of the global approach in bringing diverse data together.

3 Within-site variability in surveys

The survey approach to cotton strip assay studies in soil suggests a situation where a number of strips are placed at random on a site and the scientist is only concerned with an average measure of loss of tensile strength. A mean and standard error are initially calculated for a specific site. Inspection of the first 78 sets of data collected together for the Workshop shows a high degree of variability on many sites. Using the coefficient of variation (CV), SD/$\bar{x}$, as the measure of variability and looking at the top and bottom of strips only, Figure 1 shows the distributions of the 2 sets of CVs. The top substrips show more variability than those from the bottom substrips, and some extremely variable sites have been encountered. The CV has no use in testing or estimating, but does provide a basis for appreciating the precision possible in an experiment, thereby aiding decisions on the size of the sample and allowing for comparisons of variability.

Because of the considerable variation encountered in

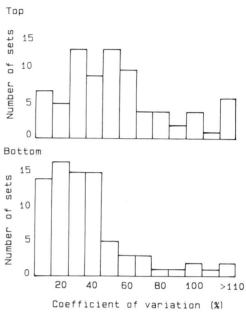

Figure 1. Histograms of coefficients of variation of top and bottom substrips of 78 decomposed cotton strips

the past on individual sites, it may be possible, in any future work, to group the sampling points into strata, in such a way that variation within a stratum can be expected to be less than variation between the strata. If successful, this method will increase the precision by which the site means can be measured.

4 Experimental design

At the start of every experiment, the question to be answered must be clearly defined. In other words, the experiment must always have a preliminary idea or a hypothesis to test. For example, the question may be: 'do the conditions of felling and not felling affect the decomposition rate of cotton strips inserted in the soil?' Some thought must also be given to the population of interest, so that the conclusions drawn are not used outside this population.

Ineson *et al.* (1988) noted that there are a large number of sets of data available about which there is knowledge of variability. This knowledge should, if possible, be used to help in the design of experiments. Is there any point in carrying out an experiment with inadequate replication in which, because of the inherent variability of the data, it is impossible to detect differences between treatments? This approach could be justified, if the experiment was planned as a pilot trial.

4.1 Example

The critical point about the importance of adequate replication can be illustrated by the following example.

It is assumed that there are 2 treatments, felling (F) and not felling (NF), and that these 2 treatments (t) have both been allocated to 3 blocks (r). A block, by definition, is a physical unit containing one complete replication of treatments. Plots within blocks must be homogeneous and differences between blocks made to account for as much as possible of the systematic variation between plots.

Five cotton strips (s) can be randomly placed in the 6 experimental plots so that the 30 cotton strips in the field form the experiment. On completion of the experiment, the following data were produced (Table 1). The grand mean is a tensile strength of 16.40. For treatment F, the mean is 19.29, and for NF it is 13.51. Can the differences between the 2 treatments be detected? Table 2 presents a skeleton analysis of variance (Steel & Torrie 1980).

Table 1. Example of a cotton strip assay data set prior to analysis

| Treatment | | Blocks | |
	I	II	III
Felling replicate	1 23.67	19.99	21.74
	2 25.28	17.17	15.25
	3 25.63	9.34	16.96
	4 10.92	19.99	22.87
	5 13.22	23.18	24.19
	Σ 98.72	89.67	101.01
	x̄ 19.74	17.93	20.2 x̄F = 19.29
Not felling replicate	1 18.67	11.89	17.90
	2 8.00	8.08	11.98
	3 7.13	14.45	14.36
	4 13.90	18.32	21.04
	5 15.60	8.25	13.13
	Σ 63.30	60.99	78.41
	x̄ 12.66	12.20	15.68 x̄NF = 13.51

Table 2. Skeleton analysis of variance

Source of variation		Degrees of freedom
Blocks	r–1	2
Treatments	t–1	1
Experimental error	(r–1) (t–1)	2
Sampling error	rt (s–1)	24
Total		29

The F ratio or variance ratio used to detect whether or not there is a significant difference is the ratio of the treatment MS/experimental error MS, in this case 250.56/4.11 = 60.92. At the 5% level, the level in the F tables is 18.51. Therefore, the result is significant. It is important to note that analysis of variance is a statistical technique for testing differences between 2 or more treatment *means*. If the analysis is reworked, based on the means, the skeleton analysis shown in

Table 3 is obtained. The degree of significance is the same.

Table 3. Skeleton analysis of variance based on means

Source of variation		Degrees of freedom
Blocks	r–1	2
Treatments	t–1	1
Experimental error	(r–1) (t–1)	2
Total	rt–1	5

The important statistical point to observe is that, by taking 5 samples within each plot, the experimental (treatment) replication has not been increased.

It would have been more satisfactory to deploy the 30 cotton strips over 5 blocks, with 3 strips allocated to each plot. In this way, the replication of the treatments has been increased from 3 to 5. The degrees of freedom for the experimental error would then rise from 2 to 4. However, it is recommended that the degrees of freedom for experimental error should be 10–20.

5 *Problems with depth and time*

When analysing the results of surveys and designed experiments using cotton strip assay, the structure of the sources of variation is often oversimplified. Simplification is usually one of 2 types: (i) analysing factorial experiments as if completely randomized, and (ii) ignoring the correlations of errors induced by sampling the same cotton strip at different depths. The first type of mistake leads to distorted probabilities in making inferences about either treatment or depth factors. The second type constitutes a serious mistake only if the correlations are not homogeneous between depths, and affects only the tests for differences between depths.

The method for analysing data which lack independence has been presented in a number of sources (Winer 1971; Gill & Hafs 1971). An appropriate analysis of variance is that for a split-plot design, as comparisons between treatments (between cotton strips) are free to vary more than comparisons between depths (within cotton strips).

For an effect lacking independence (depth or interactions), the degrees of freedom for both numerator and denominator of the global F tests (eg 'all levels are equal' *vs* 'at least 2 differ') should be multiplied by a correction factor, epsilon (Greenhouse & Geisser 1959). However, Boik (1981) has shown that the adjustment is limited to global F tests and should not be used for specific contrasts. He recommended testing each contrast against its own variance. Barcikowski and Robey (1984) recommended also using partitioned errors in *a posteriori* tests.

During the course of the Workshop, we will have seen several examples of how the cotton strip assay performs over time. Marked seasonal trends have been detected on several sites (Brown & Howson 1988; Lawson 1988). In these situations, a number of cotton strips would have been sampled at specified times throughout the season, and a site mean produced for each time. If there are several sites and the question is whether there are differences between the trends at different sites, it may be possible to subdivide the main effect of time into polynomial components and to test whether they have a significant interaction with sites.

6 *Use of the global approach in combining diverse data*

In decomposition studies of cotton strips in the tundra, Heal *et al.* (1974) successfully used the techniques of multiple regression and principal component analysis to 'explain' tensile strength losses on 24 sites. However, we have no means of knowing, from their paper, whether their predictive equation would be successful if it were applied to an independent set of data. In the intervening interval since 1974, their equations have been used on another data set (Smith & Walton 1988), and possibly also by other workers.

A major problem in combining diverse data sets is the difficulty in obtaining a complete data set. It is one of the benefits of the Workshop that an attempt can be made to construct a useful data set. Ineson *et al.* (1988) demonstrate the application of multiple regression techniques to the data from the Workshop.

If sufficient complete sets can be assembled, it would be valuable to construct predictive multiple regression models for the major geographical regions, eg tundra, tropics, temperate zone, etc. Provided that these models can be successfully assembled, then we should be able to test for differences between the regions. A useful approach to adopt would be a detailed study of the residuals, ie the difference between the observed and predicted value for each site. Extreme values, either positive or negative, could be examined in detail, and further clues might emerge to explain any large difference obtained at that site.

7 *Conclusions*

In this broad review, we draw attention to some statistical problems raised in the collection and analysis of cotton strip assay data. It has been noted that a high degree of variability can be expected when cotton strips are laid down on individual sites. Notice should be taken of this variability when organizing surveys and designing experiments, so that they are adequately replicated. Multiple regression techniques are suggested as being one of the better ways of comparing data from diverse sources, in spite of all the difficulties likely to be encountered in obtaining a complete data set.

8 *References*

Barcikowski, R.S. & Robey, R.R. 1984. Decisions in single group repeated measures analysis: statistical tests and three computer packages. *Am. Statistn,* **38,** 148-150.

Brown A.H.F. & Howson G. 1988. Changes in tensile strength loss of cotton strips with season and soil depth under 4 tree species. In: *Cotton strip assay: an index of decomposition in soils,* edited by A.F. Harrison, P.M. Latter & D.W.H. Walton, 86-89. (ITE symposium no. 24.) Grange-over-Sands: Institute of Terrestrial Ecology.

Boik, R.J. 1981. *A priori* tests in repeated measures designs: effects of nonsphericity. *Psychometrika,* **46,** 241-255.

Gill, J.L. & Hafs, H.D. 1971. Analysis of repeated measurements of animals. *J. Anim. Sci.,* **33,** 331-336.

Greenhouse, S.W. & Geisser, S. 1959. On methods in the analysis of profile data. *Psychometrika,* **24,** 95-112.

Heal, O.W., Howson, G., French, D.D. & Jeffers, J.N.R. 1974. Decomposition of cotton strips in tundra. In: *Soil organisms and decomposition in tundra,* edited by A.J. Holding, O.W. Heal, S.F. MacLean & P.W. Flanagan, 341-362. Stockholm: Tundra Biome Steering Committee.

Ineson, P., Bacon, P.J. & Lindley, D.K. 1988. Decomposition of cotton strips in soil: analysis of the world data set. In: *Cotton strip assay: an index of decomposition in soils,* edited by A.F. Harrison, P.M. Latter & D.W.H. Walton, 155-165. (ITE symposium no. 24.) Grange-over-Sands: Institute of Terrestrial Ecology.

Lawson G.J. 1988. Using the cotton strip assay to assess organic matter decomposition patterns in the mires of South Georgia. In: *Cotton strip assay: an index of decomposition in soils,* edited by A.F. Harrison, P.M. Latter & D.W.H. Walton, 134-138. (ITE symposium no. 24.) Grange-over-Sands: Institute of Terrestrial Ecology.

Smith, M.J. & Walton, D.W.H. 1988. Patterns of cellulose decomposition in four subantarctic soils. *Polar Biol.* In press.

Steel, R.G.D. & Torrie, J.H. 1980. *Principles and procedure of statistics. A biometrical approach.* New York: McGraw-Hill.

Winer, B.J. 1971. *Statistical principles in experimental design.* New York: McGraw-Hill.

The presentation of cotton strip assay results

D W H WALTON
British Antarctic Survey, Cambridge

1 Summary
The decomposition of cotton strips in the soil is measured by a change in the tensile strength (TS) of sample substrips comprised of a standardized number of cotton threads. The development of the assay is discussed in Latter and Walton (1988). A variety of techniques have been used to present such tensile strength data. This paper illustrates some of them and comments on their effectiveness.

2 Tabulated data
Tabulation of data is a common method of presentation. Older data were often obtained in pounds, but modern tensile strength testers (tensometer) give breaking strain in kilograms or newtons (N) (conversion constant 1 kg = 9.806 N).

Raw data are not easy for the reader to interpret. It is suggested that any data presented in tables should be for tensile strength lost from cotton cloth (CTSL), or tensile strength remaining, given as percentage change from initial TS (Table 1), or a control value. Table legends should state if percentages are derived from mean or median values, and give the number of replicates and the control TS value. Percentages based on means provide a good indication of the direction of change and allow immediate comparisons between different sites, but give no indication of the microsite variability in decomposition characteristic of each site.

Table 1. The mean tensile strength loss (%) of Shirley Soil Burial Test Fabric for 9 depths, in a sub-antarctic peat soil over a period of 15 months (n = 8)

Soil depth (cm)	9*	22*	43	56*	65
0–2*	10	26	34	68	65
3–5	27	59	59	85	81
6–8	36	51	65	90	85
9–11	53	60	80	95	95
12–14	71	87	92	98	98
15–17	56	85	93	100	98
18–20	52	69	88	93	93
21–23*	39	63	78	87	89
24–26	19	49	75	81	87

*Data sets used for Table 2

One method of standardizing to aid comparability is to convert the basic data to a single standard time base, eg per day, per month, per 100 days, per year. However, when the base period and standard period differ considerably, the recalculated data can be misleading because of seasonal variability. Table 2 illustrates another effect of this type of recalculation. The calculated percentage CTSL for any assay period at the

Table 2. Mean tensile strength loss (CTSL %) of Shirley Soil Burial Test Fabric calculated for 2 depths (0–2 cm and 21–23 cm)* and 3 standard time periods (per day, month and year)*, using data from 3 sample periods in a sub-antarctic peat soil

Sample period (weeks)	Measured CTSL 0–2	21–23	Calculated CTSL day 0–2	21–23	month 0–2	21–23	year 0–2	21–23
9	10	39	0.16	0.62	5.0	19.5	56.5	226.0
22	26	63	0.16	0.41	5.2	12.6	60.1	149.3
56	68	87	0.17	0.22	5.2	6.7	61.8	81.0

*Using data as indicated in Table 1

lowest soil level, based on a sampling period of 9 weeks, was 2.9 times that based on 56 weeks sampling (226–81% per year), whereas data for the upper soil level were surprisingly similar for all assay periods, ie per day 0.16–0.17%, per month 5.0–5.2%, per year 56.5–61.8%. Thus, CTSL at this site seemed to be proportional with time at the top of the soil profile, but at the base the usual curvilinear relationship applied.

3 Graphical presentation
For clarity of presentation, figures should not contain too many lines. The data in Table 1 can be shown graphically, but only about 5 of the soil profile depths can be presented in a single figure without the illustration becoming too complex (Figure 1). Where data are limited and a clear indication of mean variability is required, this type of presentation may be the best choice.

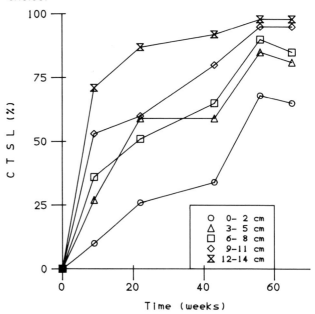

Figure 1. Mean changes in tensile strength, due to decomposition, of 2 cm wide strips of Shirley Soil Burial Test Fabric at 5 depths in a sub-antarctic peat soil (n = 8)

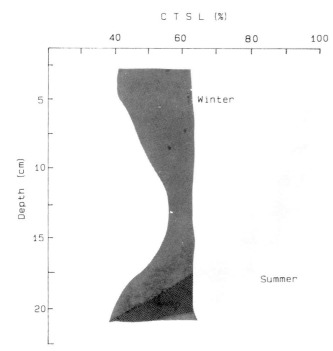

Figure 2. *Maximum and minimum values for tensile strength loss down the soil profile in a sub-antarctic peat soil, for summer and winter periods*

For some sites, it may be useful to demonstrate the overall seasonal differences only, with an indication of the variability in decay throughout the profile. Figure 2 shows the maximum and minimum CTSL through the soil profile based on seasonal assays.

Three-dimensional plots are an effective way to represent the data from consecutive assays (Figure 3).

This presentation produces a 'response surface' for a soil profile integrating depth, time and CTSL. Data for this type of figure may be derived from assays in 2 types of sequence (see Figure 1 in Lawson 1988), either with a simultaneous insertion date and several retrieval dates, at intervals of time, or with a sequence of discrete assays at various insertion dates.

4 Derived functions

To aid the comparability of data from widely different sites – a primary aim of the standard assay, several workers have suggested the use of derived functions to standardize decomposition rates. The most recent of these functions is the cotton rotting rate (CRR), obtained by a linearizing transformation (Hill *et al.* 1988). The mean value of CRR for a year is an estimate of the number of cotton strips that would be decomposed to 50% TS over a period of time (usually expressed yr^{-1}). Aberrant CRR values will be obtained if many strips have less than 10% or more than 90% CTSL. Figure 4 shows the exaggerated values for CRR obtained for very rotted strips, compared to the patterns for less extreme samples (see also Figure 1 in Brown & Howson 1988). The mean data in Figure 4 represent a single sampling on all sites, with the CRR value for the dwarf shrub site extrapolated from a period shorter than one year. However, if more samples were used to obtain data within the working limits for mean CTSL values fixed by Hill *et al.* (1988), one would then have to compare CRR values derived from multiple samples for one site with single sampling for other sites.

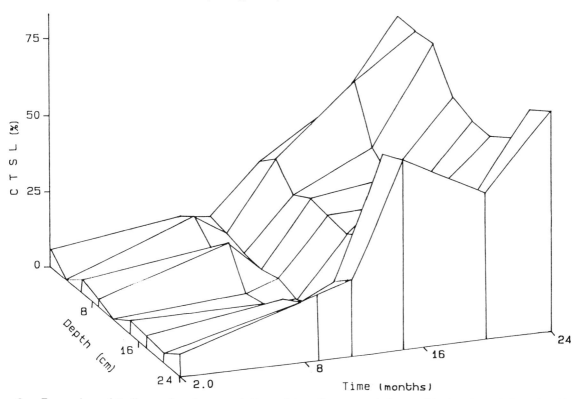

Figure 3. *Examples of 3-dimensional presentation of tensile strength loss with depth and time for (i) a sub-antarctic peat and (ii) a temperate forest soil (Brown & Howson 1988)*
i. Cumulative means for a single series of strips inserted at mid-summer (December) and retrieved at intervals

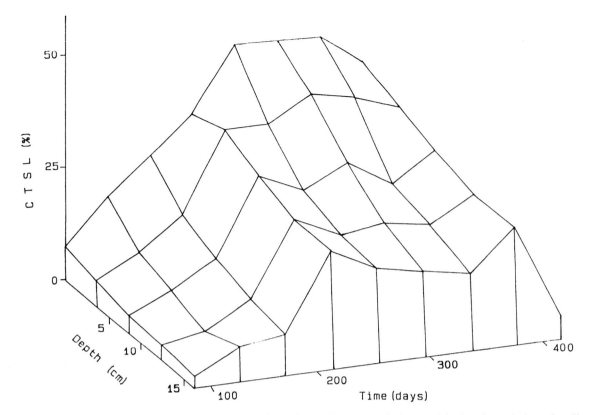

Figure 3. Examples of 3-dimensional presentation of tensile strength loss with depth and time for (i) a sub-antarctic peat and (ii) a temperate forest soil (Brown & Howson 1988)
ii. Sequential series of assays at 6-week intervals

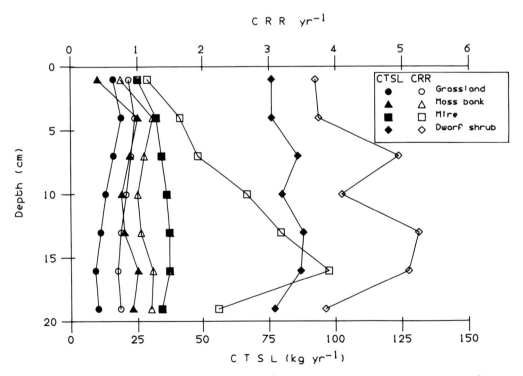

Figure 4. Tensile strength loss plotted as kg yr^{-1} and as cotton rotting rate (CRR yr^{-1}) for 4 sub-antarctic soils. Note that both mire and dwarf shrub show exaggerated CRR patterns due to the inclusion of almost completely decomposed cotton strips

5 References

Brown, A.H.F. & Howson, G. 1988. Changes in tensile strength loss of cotton strips with season and soil depth under 4 tree species. In: *Cotton strip assay: an index of decomposition in soils*, edited by A.F. Harrison, P.M. Latter & D.W.H. Walton, 86-89. (ITE symposium no. 24.) Grange-over-Sands: Institute of Terrestrial Ecology.

Hill, M.O., Latter, P.M., & Bancroft, G. 1988. Standardization of rotting rates by a linearizing transformation. In: *Cotton strip assay: an index of decomposition in soils*, edited by A.F. Harrison, P.M. Latter & D.W.H. Walton, 21-24. (ITE symposium no. 24.) Grange-over-Sands: Institute of Terrestrial Ecology.

Latter, P.M. & Walton, D.W.H. 1988. The cotton strip assay for cellulose decomposition studies in soil: history of the assay and development. In: *Cotton strip assay: an index of decomposition in soils*, edited by A.F. Harrison, P.M. Latter & D.W.H. Walton, 7-10. (ITE symposium no. 24.) Grange-over-Sands: Institute of Terrestrial Ecology.

Lawson, G.J. 1988. Using the cotton strip assay to assess organic matter decomposition patterns in the mires of South Georgia. In: *Cotton strip assay: an index of decomposition in soils*, edited by A.F. Harrison, P.M. Latter & D.W.H. Walton, 134-138. (ITE symposium no. 24.) Grange-over-Sands: Institute of Terrestrial Ecology.

The problem of cementation

D D FRENCH
Institute of Terrestrial Ecology, Banchory Research Station, Banchory

1 Summary
Under certain soil conditions, cotton strips may become 'cemented' and this may increase their measured tensile strength from its 'true' (uncemented) value. Two kinds of cementation, concretion (cementation by fine solids) and biotic cementation (resulting from microbial growth), are recognized, and some suggestions are made for dealing with the problem when comparing different soil types.

2 The problem
In some soil types, or with particular conditions or treatments, cotton strips may become 'cemented', ie some external agent acts on the fibres to bind them together, thereby possibly increasing their aggregate tensile strength (TS) over the uncemented state. Also, some of the features of fabrics listed by Sagar (1988) as having an effect on tensile strength (particularly yarn strength and effects of crossing threads) could be influenced by external agents in soil.

French (1984) distinguished 2 kinds of cementation:

i. concretion — cementing by fine solids, eg clay or lime;
ii. biotic cementation — cementing by a variety of biotically produced bonding agents, eg microbial polysaccharides and root exudates; resins may also be involved, but are often only a superficial deposit.

The first is an essentially abiotic 'physical' process, often not easily detectable without uncemented strips for comparison, but its effects can be measured by assessing the effect of artificial cementation with the relevant cementing agent (eg Figure 1). However, care

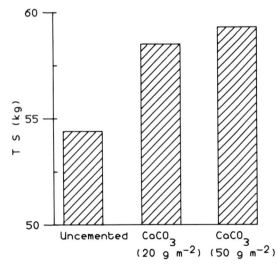

Figure 1. Effects of artificial cementation (concretion) with $CaCO_3$ on tensile strength (TS) of unrotted cotton strips. TS in both treatments is significantly higher than uncemented TS (P<0.001, t-test)

is necessary to ensure adequate mimicking of the field conditions, especially the processes involved in impregnating the cloth with the cementing agent; compare the attempts to cement cloth with calcium salts by French (1984) and Latter *et al.* (1988) using different methods. Occasionally, an experienced cotton strip handler may be able to detect from the 'feel' of a strip whether it is cemented in this way.

Biotic cementation, conversely, is usually easily detectable from the 'feel' of a strip and its fraying behaviour. Symptoms include:

i. observable coating of the threads (under a low-power light microscope);
ii. perceptible thickening of the cloth;
iii. a tendency to a felted texture;
iv. sticky or slimy coatings, especially when still wet, often hardening on drying;
v. cloth not cutting as cleanly as uncemented cloth;
vi. threads tending to stick together when test segments are frayed.

This list of indicators is not exhaustive, and all these features may not be present together, even in a heavily cemented strip, but, using these or similar criteria, it is possible at least to rank strips as uncemented, cemented or heavily cemented. However, while biotic cementation is easily detectable, it is not easy to induce it artificially under controlled conditions, especially in combination with normal activity of soil microflora. Further, because it is usually only present on partly decayed strips, it may not be possible to measure its effects on the TS of the cloth. There are some indications (Latter *et al.* 1988) that the potential effect may be at least as great as that of concretion, and Figure 2 shows that some effect was probably present in at least one field study (French 1988).

However, TS increases are not all due to cementation. The early TS increase observed by Holter (1988), for example, is more likely to be an effect of removal of lubricants, as noted by Smith and Maw (1988). The large increases in TS observed in field control strips in moss turves on Signy Island (Wynn-Williams 1988; Heal *et al.* 1974) do, however, seem to be related to the extremely high quantities of mucopolysaccharides found in those sites (D D Wynn-Williams pers. comm.), so this example may be an extreme case of biotic cementation. Latter *et al.* (1988) were not able to reproduce this effect with mucopolysaccharides alone in pure culture, but there could be many possible interactions of such potential cementing agents with other soil constituents, greatly increasing the overall degree of cementation.

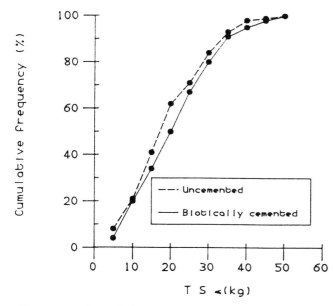

Figure 2. Cumulative percentage frequencies of TS of uncemented and detectably biotically cemented strips from a field experiment. Cemented strips have higher TS than uncemented ones (Kolmagorov–Smirnov test, n1 = 213, n2 = 207, X² = 5.36, 0.05<P<0.1). Difference in TS at 50% frequency = 3.5 kg, maximum difference = 4 kg

3 Some partial remedies

3.1 In any given study, if all strips are in the same type of soil, and nearly all strips are cemented or uncemented to a similar extent, then no correction is needed for intra-experiment comparisons, but cementation should still be noted, if present.

3.2 Where different types of soil are being compared, especially if some are likely to contain concreting agents (all clay or limestone soils so far tested do appear to cement cotton cloth) and others not (eg sandy brown earths), or where a potentially concreting soil treatment, eg liming, has been applied, then an appropriate correction factor should be derived, along the lines suggested by French (1984), and applied to all strips in the concreted batch, as either all or none will be affected.

3.3 Biotic cementation can usually be ignored if detectable only in a few strips or test segments, but, if present in more than approximately 10%, all cemented substrips should be noted. Then, either an appropriate correction should be applied *only* to substrips with detectable cementation and its effects tested on the final result (see French 1984), or the uncorrected data should be interpreted with care.

3.4 If the TS of a cemented batch of strips is significantly less than that of an uncemented batch, the difference is 'real', ie it may confidently be concluded that the cemented sample is more decayed. If TS of cemented strips is more than uncemented, a test of probable cementation effects is essential. That is, explicit corrections for cementation, appropriate to the sample, should be applied, and the relevant comparisons calculated both with and without that correction. Because even a 1% or 2% correction may make the difference between a statistically significant and a non-significant result if the variation is small enough, or the sample large enough, great care is needed in interpretation, and generally both corrected and uncorrected data should be presented.

4 References

French, D.D. 1984. The problem of 'cementation' when using cotton strips as a measure of cellulose decay in soils. *Int. Biodeterior.*, **20,** 169-172.

French, D.D. 1988. Some effects of changing soil chemistry on decomposition of plant litters and cellulose on a Scottish moor. *Oecologia.* In press.

Heal, O.W., Howson, G., French, D.D. & Jeffers, J.N.R. 1974. Decomposition of cotton strips in tundra. In: *Soil organisms and decomposition in tundra,* edited by A.J. Holding, O.W. Heal, S.F. MacLean & P.W. Flanagan, 341-362. Stockholm: Tundra Biome Steering Committee.

Holter, P. 1987. Cellulolytic activity in dung pats in relation to their disappearance rate and earthworm biomass. In: *Cotton strip assay: an index of decomposition in soils,* edited by A.F. Harrison, P.M. Latter & D.W.H. Walton, 72-77. (ITE symposium no. 24.) Grange-over-Sands: Institute of Terrestrial Ecology.

Latter, P.M., Bancroft, G. & Gillespie, J. 1988. Technical aspects of the cotton strip assay method. *Int. Biodeterior,* **24.** In press.

Sagar, B. 1988. The Shirley Soil Burial Test Fabric and tensile testing as a measure of biological breakdown of textiles. In: *Cotton strip assay: an index of decomposition in soils,* edited by A.F. Harrison, P.M. Latter & D.W.H. Walton, 11-16. (ITE symposium no. 24.) Grange-over-Sands: Institute of Terrestrial Ecology.

Smith, R.N. & Maw, J.M. 1988. Relationships between tensile strength and increase in metabolic activity on cotton strips. In: *Cotton strip assay: an index of decomposition in soils,* edited by A.F. Harrison, P.M. Latter & D.W.H. Walton, 55-59. (ITE symposium no. 24.) (Grange-over-Sands: Institute of Terrestrial Ecology.

Wynn-Williams, D.D. 1988. Cotton strip decomposition in relation to environmental factors in the maritime antarctic. In: *Cotton strip assay: an index of decomposition in soils,* edited by A.F. Harrison, P.M. Latter & D.W.H. Walton, 126-133. (ITE symposium no. 24.) Grange-over-Sands: Institute of Terrestrial Ecology.

A critical evaluation of the cotton strip assay

P J A HOWARD

Institute of Terrestrial Ecology, Merlewood Research Station, Grange-over-Sands

1 Summary

The structure and decomposition of cellulose are discussed, and the traditional definition of cellulose decomposition as the release of glucose units is upheld. The cotton strip assay is described, and it is concluded that changes in tensile strength of cotton strips cannot be related directly to specific biochemical processes which would be of interest to soil biologists.

The broader question of cellulose decomposition in soils is discussed, and the disruption to soil physiological processes caused by adding pure cellulose, in whatever form, is noted.

Published applications of the cotton strip assay are cited, and appear to be associated with poor conceptual models of soil physiological systems. It is concluded that the rate of breakdown of pure cellulose added to soil cannot provide an index of litter decomposition rate, release of litter nutrients, or 'general biological activity'.

2 Introduction

The cotton strip assay (Latter & Howson 1977; French & Howson 1982) was put into practical use before it was properly evaluated and before the relationships, if any, between tensile strength change of cotton cloth and soil processes relevant to soil research programmes were examined. Even now, some 18 years later, this situation still holds, and the significance of tensile strength changes in terms of soil processes is a matter for conjecture. The purpose of this paper is to discuss (i) the structure and biochemical decomposition of cellulose, (ii) cotton strip tensile strength changes in relation to enzymic degradation, and (iii) published applications of the cotton strip assay, and to examine the wider aspects of the measurement and importance of cellulose decomposition in soils.

Perhaps the clearest expression of the reasons for using the assay is given by Walton and Allsopp (1977): (i) cellulose is a major constituent of plant remains; (ii) the decomposition of dead plant remains is a major biological process, of great interest to many scientists studying soil processes; (iii) cellulose provides an important food source for a wide variety of organisms; and (iv) a method is needed to compare rates of breakdown of cellulose in different soils. Cotton is a natural substrate, and degradation of any material must begin with bond breaking, which leads to changes in tensile strength. Walton and Allsopp considered that, as long as this technique is used for comparative assessments of biological activity in different soils, it will remain a powerful tool for field research.

3 The structure and decomposition of cellulose

Cellulose is a homopolymer, consisting of glucose moieties joined in β-1, 4 linkages, the number of glucose moieties in a molecule being the degree of polymerization (DP). Native cellulose from higher plants has a DP of about 14 000, and this value appears to be remarkably constant. X-ray and infra-red data suggest that the basic structure of cellulose involves the anhydrocellobiose unit ($C_{12}H_{20}O_{10}$) rather than the anhydroglucose unit ($C_6H_{10}O_5$), so that the shortest cellulose-like molecule would be cellotetrose (DP = 4):

$$(C_6H_{11}O_5) . (C_{12}H_{20}O_{10}) . (C_6H_{11}O_6)$$

Here, ($C_6H_{11}O_5$) represents a glucose moiety with a free secondary hydroxyl group on C-4, and ($C_6H_{11}O_6$) is a glucose molecule with a reducing group. Cellobiose, cellotriose, and cellotetrose are water-soluble, cellopentose is very sparingly soluble, and cellulose molecules with DP greater than 6 are insoluble (Ljungdahl & Eriksson 1985).

Because cellulose is a polymer, we have to think rather carefully about how we define cellulose decomposition. In nature, cellulose is degraded by a range of aerobic and anaerobic fungi and bacteria, many of which occur in extreme conditions of temperature and pH. A list of fungi which have been examined in cellulose decomposition studies is given in Ljungdahl and Eriksson (1985). Many fungi can utilize oligosaccharides and polysaccharides as carbon sources, but, in general, these molecules are too large to be transported directly into the cell, and must first be hydrolyzed to their subunits. Cellulose represents an important potential carbon and energy source for fungi, many of which produce a series of enzymes, collectively called cellulase (Sagar 1988), which facilitate the degradation of cellulose to glucose units.

Cellulose-degrading enzyme systems have been studied in detail in 2 fungi, *Sporotrichum pulverulentum*, the conidial state of the white-rot fungus (*Phanerochaeate chrysosporium*) (Eriksson 1981), and the mould *Trichoderma reesei* (Ryu & Mandels 1980). They have similar hydrolytic enzyme systems with at least 3 components:

i. endo-1, 4-β-glucanases, which split randomly 1, 4-β-glucosidic linkages within the cellulose polymer;
ii. exo-1, 4-β-glucanases, which split off either cellobiose or glucose from the non-reducing end of the cellulose polymer;
iii. 1, 4-β-glucosidases, which hydrolyze cellobiose and water-soluble cellodextrins to glucose.

However, relatively little plant cellulose occurs in the more or less pure form, for much of it is encrusted by lignin, hemicelluloses, and pectins. Some white-rot basidiomycetes selectively remove lignin and hemicellulose without degrading much cellulose, whilst others remove all the cell wall components (Otjen & Blanchette 1985).

Brown-rot fungi are generally basidiomycetes and decompose mainly wood polysaccharides. They produce component (i) but lack (ii) (Highley 1975a, b), and the biochemistry of cellulose degradation by them is not fully understood. It has been found that brown-rot fungi can attack cellulose at some distance from the fungal cell wall, and the initial attack could be by low molecular weight substances that diffuse easily through wood cell walls. Brown-rot fungi degrade cellulose in wood, but they do not seem to degrade pure, isolated cellulose (Nilsson 1974a; Highley 1977). Highley (1978) found that, in culture, the brown-rot fungus (*Poria placenta*) could not digest crystalline cellulose, unless glucose, starch, holocellulose, or mannose was added. Hydrolysis of crystalline cellulose is thought to occur with endo-glucanases attacking randomly over the cellulose polymer, creating end groups for the exo-glucanases to attack from the non-reducing ends of the chains. Degradation of highly ordered cellulose in native cotton fibres depends almost completely on the synergistic action of the 3 enzyme systems listed above (Sagar 1988). None of them acting independently can solubilize cotton to a significant extent (Eriksson & Wood 1984).

Bacterial cellulase systems appear to be more complex than those of fungi, and they cannot be characterized neatly in terms of endo-1, 4-β-glucanases and exo-1, 4-β-glucanases. Many bacteria do not possess β-glucosidase (cellobiase) or, if they possess it, they may still metabolize cellodextrins and cellobiose by phosphorylases (Ljungdahl & Eriksson 1985). In the end, in aerobic conditions, the result is the formation of glucose.

So, how do we define cellulose decomposition? A reduction in the DP, brought about by the action of the endo-glucanases, results in molecules which are still, by definition, cellulose, until the DP falls to a very small value. Cellulose molecules with a DP greater than 4 are very sparingly soluble or insoluble. Many micro-organisms can use glucose, and the reduction of the cellulose molecule to its basic glucose units is the logical definition of cellulose decomposition. This definition has been used by soil researchers for many years, and cellulose decomposition has long been measured by glucose production.

4 The relationships of cotton strip tensile strength changes to soil processes

The structure of cotton fibres is complex (Sagar 1988). It seems to be generally agreed that the glucose units are linked together to give a somewhat kinked but rather rigid chain, about 2000 nm long and 0.75 nm wide. However, this structure alone is not sufficient to account for the physical properties of cellulose fibres (Rogers 1961). The strength of cotton fibres depends on many structural features, such as the average chain lengths of the molecules, their homogeneity, and orientation. A close correlation has been found between the strength of a fibre and the degree of polymerization of the component cellulose molecules (Siu 1951). Bundles of chains are thought to be held together by hydrogen bonds and van der Waal's forces, to form microfibrils about 5 nm in diameter. Few, if any, cellulose chains cross from one microfibril to another, and the microfibrils are built up into layers (Selby 1968).

The mode of breakdown of cotton fibres by soil organisms is not fully understood. Work cited in Section 3 suggests that enzymic attack by endo-glucanases would progressively reduce the DP, and the synergistic effect of these, the exo-glucanases, and the glucosidases would lead eventually to glucose. Random severing of the glycosidic linkages of the cellulose polymer would bring about a reduction in both the chain length of the molecules and their homogeneity. In this way, there would be an initial fall in tensile strength, without much loss of glucose units. The relationship between tensile strength and weight loss suggests such an effect (Siu 1951; Selby 1968; Heal *et al.* 1974). There is also evidence that the inaccessibility of cotton fibres to large protein molecules delays enzymic attack and results in the initial breakdown being localized (Sagar 1988). Nilsson (1974b) found that 15 out of 20 fungal species tested could produce cavities in cotton fibres, and 18 species could produce surface erosion. Localized effects of this type will reduce tensile strength with small weight loss.

That such changes do actually occur was shown by Halliwell (1965), who found that enzymic degradation of purified cotton fibres resulted in the release of insoluble, non-filter-passing, very short fibres; insoluble, filter-passing products; and soluble products, mainly glucose.

It has been suggested that cotton strip tensile strength changes might be calibrated in terms of weight loss. The results of Halliwell show that weight loss itself is not readily interpretable, as much of it represents products of partial decomposition. Ghewande (1977) studied loss in weight of filter paper and of cotton in cultures of 5 plant pathogenic fungi. For all the fungi, loss in weight of filter paper was greater than that of cotton, and the ratios of the weight losses differed for the different fungi. Different artificially processed cellulose materials seem to be attacked in different ways. In general, there was no correlation between loss in weight of the cellulose and cellulase production by a fungus (Ghewande & Deshpande 1975).

5   Application of the cotton strip assay

5.1   Its use in ecological studies

Rovira (1953) measured changes in tensile strength of cotton tapes buried in soil, but his method has not been much used. The first recorded ecological use of the assay, as described by Latter and Howson (1977), appears to have been by Springett (1971), for assessing 'relative decomposition rates' in respect to maritime pine (*Pinus pinaster*) litter. Latter and Howson (1977) described the method as used in the tundra biome of the International Biological Programme, for assessing cellulose decomposer activity (Baker 1974; Heal *et al.* 1978). Other users have been Wynn-Williams (1979, 1980) to study cellulose decomposition in antarctic soils; Miles (1981) and Miles and Young (1980) to study cellulose decomposition in heathland and moorland soils colonized by birch (*Betula* spp.); and Brown *et al.* (1983), Brown (1988) and Brown and Howson (1988) to assess the 'potential for decomposer activity' in soils under alder (*Alnus glutinosa*), Scots pine (*Pinus sylvestris*), oak (*Quercus* spp.), and Norway spruce (*Picea abies*).

5.2   Difficulties in interpretation of results

Several points which have arisen in various papers indicate that there are difficulties in interpretation, in terms of soil organic matter decomposition processes.

Heal *et al.* (1974) and Latter and Howson (1977) have computed a series of regressions relating change in tensile strength of cotton strips to weight loss of cotton following decomposition. Though the relationships were statistically significant, in some cases highly significant, there were also very significant differences in the regressions for the different sites studied. The authors themselves state that 'a strict comparison between TS and weight losses cannot be expected; weight loss represents an average measure of decomposition over the whole test piece, whereas the tensile strength is a measure of loss at the most decomposed part'.

There appears to be no correlation between the decomposition of cellulose and cellulase activity in soils (Ross *et al.* 1978; Ross & Speir 1979).

Heal *et al.* (1974) used the cotton strip assay, essentially as described in Latter and Howson (1977), for relating decomposition potential to weight losses of litter. They found that (i) with low weight loss of cotton strips, there was a very rapid loss of tensile strength, (ii) in very wet conditions, there was a decline in tensile strength loss of cotton strips, although litters continued to show high weight losses under the same conditions, and (iii) there was an important effect of litter quality in the processes of decomposition.

5.3   Untenable arguments put forward to justify use of the assay

5.3.1   Cellulose is a major constituent of plant detritus

The fact that cellulose is the major constituent of plant remains does not justify the use of cotton strips, or any other pure and, especially, processed cellulose substrate, to simulate the decomposition of cellulose in plant remains. Plant cellulose does not enter the soil in pure form; most plant cellulose is encrusted with lignin and hemicelluloses. Plant material added to soil in natural conditions always contains some nitrogen (N), and materials from various origins have characteristic carbon/nitrogen (C/N) values. Populations of soil organisms develop during the degradation of these materials, and the resulting soil organic matter also has characteristic C/N values. The response of the soil biota to the pure cellulose substrate will necessarily differ from that to plant litters. If anything is needed in such studies, it is a method for studying the decomposition of cellulose, and other constituents, in plant litter.

5.3.2   The decomposition of dead plant detritus is a biological process of considerable importance in maintaining soil fertility

The idea that cellulose decomposition can represent litter decomposition adequately is clearly a gross oversimplification. Minderman (1968) showed that different litter constituents decompose at different rates, and accumulation of residues is determined mainly by the content of intractable constituents. Various authors have preferred to use lignin content as a predictor of litter decomposition rate (Fogel & Cromack 1977; Meentemeyer 1978; Melillo *et al.* 1982). Changes in log biomass were also correlated with lignin content (Charley & Richards 1983). Berg and Staaf (1980) found that, in the early stages of decomposition of Scots pine needle litter, ie up to 30% weight loss, loss in weight was due mainly to soluble substances and degradation of cellulose and hemicellulose and was correlated with the initial levels of N, phosphorus (P), potassium (K) and sulphur (S). Decomposition of lignin was unaffected by initial nutrient levels, and its decomposition started sooner in litters with much lignin than in those with little. In the later stages of litter decomposition, weight loss depended on lignin decomposition.

It is a matter of common observation that litters of different tree species decompose at different rates on the same soil type, and that litter of a given species decomposes at different rates on different soil types. The rate of litter decomposition is not a property solely of the litter, but is the result of a complex interaction between plant species and soil chemistry (eg Coulson *et al.* 1960; Davies *et al.* 1964). Similar effects appear to occur in grasslands, though perhaps they are less pronounced (eg Barratt 1965).

5.3.3   Cellulose provides an important food source for a wide variety of organisms

Cellulose is certainly an important carbon and energy source for micro-organisms, but polysaccharide molecules are too large to be transported into cells, and cellulose must first be broken down to glucose units.

Because of the structure of cotton fibres and the mode of action of cellulases on them, tensile strength change cannot be related to glucose release.

The degradation of any (purified cellulose) material must begin with bond breaking, leading to changes in its tensile strength, but this has more to do with the properties of textiles than with the study of the transformation of plant remains in soils. The importance of cellulose breakdown in soil systems lies in the fact that it provides a source of carbon and energy for a range of soil organisms, but they can only use the glucose formed. Changes in the tensile strength of cotton strips are closely related to the DP, which is reduced by the bond breaking action of the endoglucanases. However, the resulting material will still be cellulose, and needs to be broken down into the constituent glucose units before it can be used by the micro-organisms. As this stage cannot be related to tensile strength changes, the latter do not provide information which is of use to soil biologists.

An accurate method for determining cellulose decomposition in soils, by measuring the release of glucose units, has existed for some years (Benefield 1971) and is discussed by Ross (1974). Broadly, the method involves incubating cellulose powder with a known quantity of soil in a buffer solution containing Penicillin G as a bacteriostat. After incubation, the glucose formed is dehydrogenated by glucose oxidase, and the hydrogen peroxide so formed is catalyzed by peroxidase to oxidize o-tolidine dihydrochloride. The latter reaction is determined colorimetrically. Penicillin G is thought not to suppress fungal activity,

so the method measures the activity of cellulase enzymes present initially in the soil sample, plus any which were synthesized during the incubation. There are important physiological differences between the cotton strip assay and the normal methods for determining cellulase activity. Although cellulase enzymes are inducible, cellulase methods such as that of Benefield (1971) use a short incubation time, which is unlikely to allow major changes in the quantity of cellulase to occur. Therefore, we may expect such a method to reveal the current propensity of the soil to decompose cellulose.

Figure 1 shows the mean cellulase activities (oven dried (OD) basis), determined by the method of Benefield (1971), of 48 soils (0–5 cm) in and around the English Lake District, sampled at 4-weekly intervals for 56 weeks. Each point in the Figure is the mean of 14 samples (Howard & Howard 1985). Figure 2 gives the same results for loss-on-ignition (LOI) basis. These data make the results for the different soils more comparable as they vary considerably in organic matter content. At any given pH value above pH 4, there is clearly a considerable range of cellulase activity; below pH 4, the range is smaller. However, these results are for soils at 0–5 cm depth, and we have to consider the meaning of cellulase activity at that depth in soils of different pH. Above pH 5, we are dealing with mull soils. Litter deposited on the soil surface does not stay there for long, but is pulled into the soil by the large earthworms (Lumbricidae). Thus, most of the litter cellulose decomposes in the soil. At the same time, fine roots are being formed and decomposed there also. In such soils, we would expect a

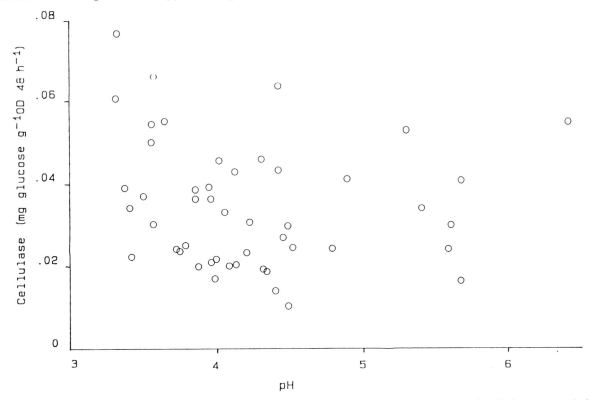

Figure 1. Plot of cellulase activity (OD basis) on pH. Means of 14 4-weekly samples of soils in 48 woods in and around the English Lake District. r = –0.137 NS, r² = 0.019

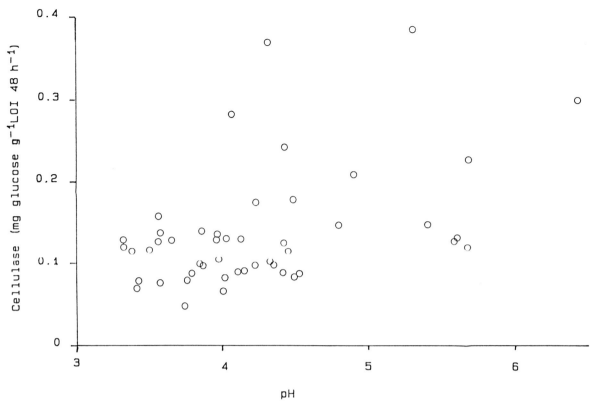

Figure 2. Plot of cellulase activity (LOI basis) on pH. 48 soils as in Figure 1. r = 0.457 P<0.01, r² = 0.209

generally high level of cellulase activity. By contrast, below pH 4, mixing of litter with mineral soil does not occur, and the litter cellulose decomposes in the superficial organic layers. Thus, cellulase activity in the surface mineral soil must be largely associated with root decomposition.

This is an important difference in cellulose decomposition in these extreme soil types, and must be recognized when interpreting the results of cellulase activity determinations. The reasons for the wide range in cellulase activities of the soils with pH greater than 5 (Figure 2) may, therefore, lie in differences in quantities of leaf and root material added to the soil. The differences in soils with pH less than 4 may be due to differences in root turnover.

By contrast with cellulase activity measurements, cotton strips are inserted into the soil in the field and left in position for weeks, which allows time for major changes in populations of organisms, and their biochemistry. The capacity of a soil to decompose cellulose is not a fixed property, like the capacity of a cow to eat hay. If we add more cellulose to a soil, the soil organism population will grow to deal with it, but this takes time. As cellulose contains no N, we would expect the addition of pure cellulose to put a severe N stress on the soil. Mull soils with high levels of N mineralization could cope with such stress better than mor soils with low levels of N mineralization. However, the real situation is not so simple, as most cellulolytic fungi show greatly reduced production of cellulases as substrate C/N increases. The white-rot fungi are exceptional in this regard, being able to

produce cellulases at a C/N of 2000 (Charley & Richards 1983, p34). It is clear that adding pure cellulose to soils stresses the physiology in ways which are largely unpredictable, making the results difficult, if not impossible, to interpret without a considerable amount of additional information. Cotton strips are likely to be decomposed in soil by a selected microflora, as indicated by Widden *et al.* (1986), and the results obtained by the assay may show little comparison with normal decomposition of plant litter or soil organic matter.

5.4 Its use as an index of 'biological activity'
It has been suggested that cotton strips may be used to provide an index of 'biological activity' of soils, eg when comparing soils under different vegetation types. The idea that the rich diversity of soil physiological activities can be represented adequately by any single measurement can be dismissed by available evidence. Plots of cellulase activity against coefficient of humidity (moisture content g^{-1} LOI), oxygen uptake, and phosphatase activity (both on LOI basis) for the 48 soils in Figures 1 and 2 are shown in Figures 3–5. The largest correlation coefficient is cellulase and pH ($r^2 = 0.209$).

When the mean values of the soils from the 48 woods were divided into 3 pH groups, <3.8, 3.8–5.0, and >5.0 (Pearsall 1938, 1952), there was no significant difference between the mean values of the groups with regard to humidity. However, for oxygen uptake (LOI basis), the mean of the pH <3.8 group was significantly lower than that of the intermediate group (19.9, 23.8, 23.6 respectively). For cellulase activity,

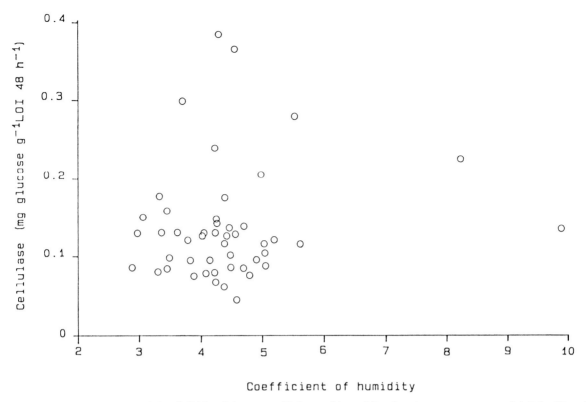

Figure 3. Plot of cellulase activity (LOI basis) on coefficient of humidity (moisture content g^{-1} LOI). 48 soils as in Figure 1. r = 0.122 NS, r^2 = 0.015

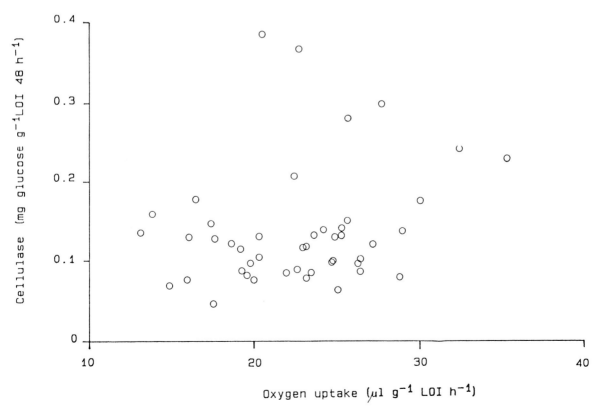

Figure 4. Plot of cellulase activity on oxygen uptake (both on LOI basis). 48 soils as in Figure 1. r = 0.226 NS, r^2 = 0.051

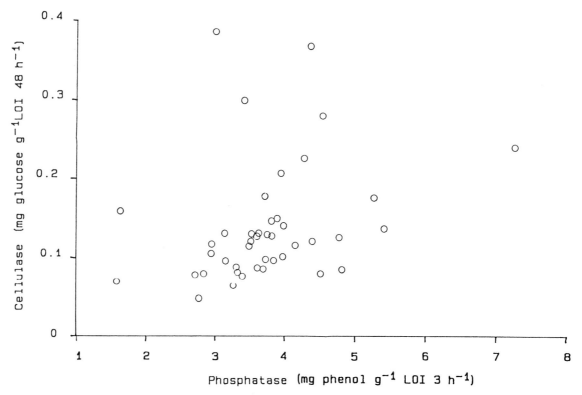

Figure 5. Plot of cellulase activity on phosphatase activity (both on LOI basis). 48 soils as in Figure 1.
r = 0.280 NS, r² = 0.078

the mean of the more acidic group and the mean of the intermediate group were significantly lower than the mean of the pH >5.0 group (0.106, 0.136, 0.207 respectively). For phosphatase activity, the mean of the more acidic group was significantly lower than that of the intermediate group (3048, 4047, 3757 respectively).

Studying enzymes extracted from coniferous leaf litter, Spalding (1977, 1980) found that cellulase activity was poorly correlated with invertase (r = 0.33, r² = 0.11), β-glucosidase (r = 0.37, r² = 0.14) and polyphenoloxidase (r = 0.29, r² = 0.08). Other workers have found correlations among soil physiological activities to be good in some cases but not in others (Ladd & Butler 1972; Hankin *et al.* 1974; Voets *et al.* 1975; Nannipieri *et al.* 1978, 1979; Frankenberger & Dick 1983). Furthermore, significant correlations between activities observed in one year may not be found in another (Hersman & Temple 1979).

Figure 2 shows that similar cellulase values are possible in the pH range 4–6. Soils at the extremes of this range have greatly different chemical and biological properties. Also, at any given pH, a wide range of cellulase activities is possible. Figures 4 and 5 show that a range of values for oxygen uptake and phosphatase activity is possible for a given cellulase activity, and a range of cellulase activities is possible for given values of oxygen uptake and phosphatase activity. It seems clear that the fact that soils have different cellulolytic activities cannot be used to make deductions about their other physiological properties. Even

if the cotton strip assay gave an adequate measure of cellulose breakdown, its use in studies such as those of Brown and Howson (1988) would be pointless.

5.5   Its use in nutrient cycling studies
It has been suggested that the cotton strip assay may be used to indicate differences in nutrient cycling under different soil/vegetation systems, for example in ITE's studies of the Gisburn Forest (Brown 1988) and of grasslands on Plynlimon, central Wales (G Howson pers. comm.). However, no plant nutrients are bound to cellulose, and none would be released by cellulose decomposition. K occurs mainly in vacuolar sap, calcium (Ca) and magnesium (Mg) occur in pectates, and Mg also occurs in chlorophyll and enzyme systems. K and Ca accumulate in stemwood, while P and N occur mostly in the nucleus and cytoplasm.

Brown and Howson (1988) present results from Gisburn Forest which show that changes in tensile strength of cotton strips reflect the growth of trees, either pure or in mixtures. As the tensile strength changes cannot be interpreted directly, a separate study is required on the factors affecting the rate of rotting of the cotton strips, and to investigate if those factors could account for the observed differences in tree growth. It would seem to be simpler to leave out the intermediate step, and to study directly the factors affecting tree growth. A simple explanation for these results would be that the rate of rotting of cotton strips and the growth of trees have nothing in common, except that they both depend on the amount of available soil N (see Carlyle & Malcolm 1986a, b).

6 Conclusions

Most published uses of cotton strips are somewhat vague about what the method is intended to show. Phrases such as 'relative decomposition rates' or 'potential for decomposer activity' are common. In general, the assumption seems to be that changes in tensile strength of cotton strips are in some way related to plant litter decomposition.

The use of pure processed cellulose as a surrogate for plant litter is based on highly oversimplified conceptual models of soil physiological systems. Such models stress quantitative aspects, while ignoring important qualitative differences (cf Romell 1935).

Furthermore, plant cellulose is not associated with nutrient elements, and its decomposition is not directly relevant to nutrient cycling. The rate of breakdown of pure cellulose is not strongly related to many other important soil physiological processes, so it cannot be used as an index of biological activity.

In my opinion, the use of an unsuitable method for assessing cellulose decomposition, in association with poor conceptual models, is unlikely to increase our knowledge of the functioning of soil biological systems, or our ability to predict the consequences of natural or artificial perturbation.

7 References

Baker, J.H. 1974. Comparison of the microbiology of four soils in Finnish Lapland. *Oikos,* **25,** 209-215.

Barratt, B.C. 1965. Decomposition of grass litters in three kinds of soil. *Pl. Soil,* **23,** 265-269.

Benefield, C.B. 1971. A rapid method for measuring cellulase activity in soils. *Soil Biol. Biochem.,* **3,** 325-329.

Berg, B. & Staaf, H. 1980. Decomposition rate and chemical changes of Scots pine needle litter. II. Influence of chemical composition. In: *Structure and function of northern coniferous forests - an ecosystem study,* edited by T. Persson, 373-390. (Ecological Bulletin 32.) Stockholm: Swedish Natural Science Research Council.

Brown, A.H.F. 1988. Discrimination between the effects on soils of 4 tree species in pure and mixed stands using cotton strip assay. In: *Cotton strip assay: an index of decomposition in soils,* edited by A.F. Harrison, P.M. Latter & D.W.H. Walton, 80-85. (ITE symposium no. 24.) Grange-over-Sands: Institute of Terrestrial Ecology.

Brown, A.H.F. & Howson, G. 1988. Changes in tensile strength loss of cotton strips with season and soil depth under 4 tree species. In: *Cotton strip assay: an index of decomposition in soils,* edited by A.F. Harrison, P.M. Latter & D.W.H. Walton, 86-89. (ITE symposium no. 24.) Grange-over-Sands: Institute of Terrestrial Ecology.

Brown, A.H.F., Gardener, C.L. & Howson, G. 1983. Differences in some biological attributes of soils developed under four tree species. In: *Trans. biological processes & soil fertility,* 58. International Society of Soil Science and British Society of Soil Science.

Carlyle, J.C. & Malcolm, D.C. 1986a. Nitrogen availability beneath pure spruce and mixed larch + spruce stands growing on a deep peat. I. Net N mineralization measured by field and laboratory incubations. *Pl. Soil,* **93,** 95-113.

Carlyle, J.C. & Malcolm, D.C. 1986b. Nitrogen availability beneath pure spruce and mixed larch + spruce stands growing on a deep peat. II. A comparison of N availability as measured by plant uptake and long-term laboratory incubations. *Pl. Soil,* **93,** 115-122.

Charley, J.L. & Richards, B.N. 1983. Nutrient allocation in plant communities: Mineral cycling in terrestrial ecosystems. In: *Physiological plant ecology IV,* edited by O.L. Lange, P.S. Nobel, C.B. Osmond & H. Zieger, 5-45. Berlin: Springer.

Coulson, C.B., Davies, R.I. & Lewis, D.A. 1960. Polyphenols in plant, humus, and soil. I. Polyphenols of leaves, litter and superficial humus from mull and mor sites. *J. Soil Sci.,* **11,** 30-44.

Davies, R.I., Coulson, C.B. & Lewis, D.A. 1964. Polyphenols in plant, humus, and soil. III. Stabilization of gelatin by polyphenol tanning. *J. Soil Sci.,* **15,** 299-309.

Eriksson, K.E. 1981. Cellulases of fungi In: *Trends in the biology of fermentations for fuels and chemicals,* edited by A. Hollaender, 19-32. New York: Plenum Press.

Eriksson, K.E. & Wood, T.M. 1984. Biodegradation of cellulose. In: *Biosynthesis and biodegradation of wood components,* edited by T. Higuchi, 469-503. New York: Academic Press.

Fogel, R. & Cromack, K. 1977. Effect of habitat and substrate quality on Douglas fir litter decomposition in western Oregon. *Can. J. Bot.,* **55,** 1632-1640.

Frankenberger, W.T. & Dick, W.A. 1983. Relationships between enzyme activities and microbial growth and activity indices in soil. *J. Soil Sci. Soc. Am.,* **47,** 945-951.

French, D.D. & Howson, G. 1982. Cellulose decay rates measured by a modified cotton strip method. *Soil Biol. Biochem.,* **14,** 311-312.

Ghewande, M.P. 1977. Decomposition of cellulose and the production of cellulolytic enzymes by plant pathogenic fungi *J. Biol. Sci.,* **20,** 69-73.

Ghewande, M.P. & Deshpande, K.B. 1975. Cellulolytic enzymes of *Helminthosporium apattarnae. Indian J. Microbiol.,* **15,** 43-45.

Halliwell, G. 1965. Hydrolysis of fibrous cotton and reprecipitated cellulose by cellulolytic enzymes from soil micro-organisms. *Biochem. J.,* **95,** 270-281.

Hankin, L., Sands, D.C. & Hill, D.E. 1974. Relation of land use to some degradative enzymatic activities of soil bacteria. *Soil Sci.,* **118,** 38-44.

Heal, O.W., Howson, G., French, D.D. & Jeffers, J.N.R. 1974. Decomposition of cotton strips in tundra. In: *Soil organisms and decomposition in tundra,* edited by A.J. Holding, O.W. Heal, S.F. MacLean & P.W. Flanagan, 341-362. Stockholm: Tundra Biome Steering Committee.

Heal, O.W., Latter, P.M. & Howson, G. 1978. A study of the rates of decomposition of organic matter. In: *Production ecology of British moors and montane grasslands,* edited by O.W. Heal & D.F. Perkins, 136-159. Berlin: Springer.

Hersman, L.E. & Temple, K.L. 1979. Comparison of ATP, phosphatase, pectinolase and respiration as indicators of microbial activity in reclaimed coal strip mine spoils. *Soil Sci.,* **127,** 70-73.

Highley, T.L. 1975a. Can wood-rot fungi degrade cellulose without other wood constituents? *Forest Prod. J.,* **25,** 38-39.

Highley, T.L. 1975b. Properties of cellulases of two brown-rot fungi and two white-rot fungi. *Wood Fiber,* **6,** 275-281.

Highley, T.L. 1977. Requirements for cellulose degradation by a brown-rot fungus. *Mater. Org.,* **12,** 25-36.

Highley, T.L. 1978. Degradation of cellulose by culture filtrates of *Poria placenta. Mater. Org.,* **12,** 161-174.

Howard, P.J.A. & Howard, D.M. 1985. *Multivariate analysis of soil physiological data.* (Merlewood research and development paper no. 105.) Grange-over-Sands: Institute of Terrestrial Ecology.

Ladd, J.N. & Butler, J.H.A. 1972. Short-term assays of soil proteolytic enzyme activities using proteins and dipeptide derivatives as substrates. *Soil Biol. Biochem.,* **4,** 19-30.

Latter, P.M. & Howson, G. 1977. The use of cotton strips to indicate cellulose decomposition in the field. *Pedobiologia,* **17,** 145-155.

Ljungdahl, L.G. & Eriksson, K.E. 1985. Ecology of microbial cellulose degradation. *Adv. microb. Ecol.,* **8,** 237-299.

Meentemeyer, V. 1978. Macroclimate and lignin control of litter decomposition rates. *Ecology,* **59,** 464-472.

Melillo, J.M., Aber, J.D. & Muratore, J.F. 1982. Nitrogen and lignin control of hardwood leaf litter decomposition dynamics. *Ecology,* **63,** 621-626.

Miles, J. 1981. *Effect of birch on moorlands.* Cambridge: Institute of Terrestrial Ecology.

Miles, J. & Young, W.F. 1980. The effects on heathland and moorland soils in Scotland and northern England following colonization by birch (*Betula* spp.). *Bull. Ecol.,* **11,** 233-242.

Minderman, G. 1968. Addition, decomposition and accumulation of organic matter in forests. *J. Ecol.*, **56,** 355-362.

Nannipieri, P., Johnson, R.L. & Paul, E.A. 1978. Criteria for measurement of microbial growth and activity in soil. *Soil Biol. Biochem.*, **10,** 223-229.

Nannipieri, P., Pedrazzini, F., Arcara, P.G. & Piovanelli, C. 1979. Changes in amino acids, enzyme activities, and biomasses during soil microbial growth. *Soil Sci.*, **127,** 26-34.

Nilsson, T. 1974a. Comparative study of the cellulolytic activity of white-rot and brown-rot fungi. *Mater. Org.*, **9,** 173-198.

Nilsson, T. 1974b. Microscopic studies on the degradation of cellophane and various cellulosic fibres by wood-attacking microfungi. *Stud. for. suec.*, no. 117.

Otjen, L. & Blanchette, R.A. 1985. Selective delignification of aspen wood blocks *in vitro* by three white-rot basidiomycetes. *Appl. environ. Microbiol.*, **50,** 568-572.

Pearsall, W.H. 1938. The soil complex in relation to plant communities. II. Characteristic woodland soils. *J. Ecol.*, **26,** 194-209.

Pearsall, W.H. 1952. The pH of natural soils and its ecological significance. *J. Soil Sci.*, **3,** 41-51.

Rogers, H.J. 1961. The dissimilation of high molecular weight substances. In: *The bacteria. Vol. 2: Metabolism,* edited by I.C. Gunsalius & R.Y. Stanier, 257-318. New York; London: Academic Press.

Romell, L.G. 1935. Ecological problems of the humus layer in the forest. *Mem. Cornell Univ. agric. Exp. Stn,* no. 170.

Ross, D.J. 1974. Glucose oxidase activity in soil and its possible interference in assays of cellulase activity. *Soil Biol. Biochem.*, **6,** 303-306.

Ross, D.J. & Speir, T.W. 1979. Studies on a climosequence of soils in tussock grasslands 23. Cellulase and hemicellulase activities of topsoils and tussock plant materials. *N.Z. Jl Sci.*, **22,** 25-33.

Ross, D.J., Molloy, L.F., Bridger, B.A. & Cairns, A. 1978. Studies on a climosequence of soils in tussock grasslands 20. Decomposition of cellulose on the soil surface and in the topsoil. *N.Z. Jl Sci.*, **21,** 459-65.

Rovira, A.D. 1953. A study of the decomposition of organic matter in red soils of the Lismore district. *Aust. Conf. Soil Sci., Adelaide,* **1,** 3.17, 1-4.

Ryu, D.D.Y. & Mandels, M. 1980. Cellulase: biosynthesis and applications. *Enzyme microb. Technol.*, **2,** 91-102.

Sagar, B.F. 1988. Microbial cellulases and their action on cotton fibres. In: *Cotton strip assay: an index of decomposition in soils,* edited by A.F. Harrison, P.M. Latter & D.W.H. Walton, 17-20. (ITE symposium no. 24.) Grange-over-Sands: Institute of Terrestrial Ecology.

Selby, K. 1968. Mechanism of biodegradation of cellulose. In: *Biodeterioration of materials,* edited by A.H. Walters & J.J. Elphick, 62-78. London; New York: Elsevier.

Siu, R.G.H. 1951. *Microbial decomposition of cellulose.* New York: Reinhold.

Spalding, B.P. 1977. Enzymatic activities related to the decomposition of coniferous leaf litter. *J. Soil Sci. Soc. Am.*, **41,** 622-627.

Spalding, B.P. 1980. Enzymatic activities in coniferous leaf litter. *J. Soil Sci. Soc. Am.*, **44,** 760-764.

Springett, J.A. 1971. The effects of fire on litter decomposition and on the soil fauna in a *Pinus pinaster* plantation. In: *4th Colloquium Pedobiologiae, Dijon, 1970,* 529-535. Paris: Institut National de la Recherche Agronomique.

Voets, J.P., Agrianto, G. & Verstraete, W. 1975. Etude ecologique des activites microbiologiques et enzymatiques des sols dans une foret de feuillus. *Revue Ecol. Biol. Sol,* **12,** 543-555.

Walton, D.W.H. & Allsopp, D. 1977. A new test cloth for soil burial trials and other studies on cellulose decomposition. *Int. Biodeterior. Bull.,* **13,** 112-115.

Widden P., Howson, G. & French, D.D. 1986. Use of cotton strips to relate fungal community structure to cellulose decomposition rates in the field. *Soil Biol. Biochem.*, **18,** 335-337.

Wynn-Williams, D.D. 1979. Techniques used for studying terrestrial microbial ecology in the maritime Antarctic. In: *Cold-tolerant microbes in spoilage and the environment,* edited by A.D. Russell & D. Fuller, 67-81. London: Academic Press.

Wynn-Williams, D.D. 1980. Seasonal fluctuations in microbial activity in Antarctic moss peat. *Biol. J. Linn. Soc.*, **14,** 11-28.

Problems and advantages of using the cotton strip assay in polar and tundra sites

D W H WALTON
British Antarctic Survey, Cambridge

1 Summary

The principal difficulties of using the cotton strip assay in tundra and polar soils are summarized. Frozen ground, rocky soils and bird interference all affect the planning of experiments and the likelihood of retrieving an adequate number of inserted strips. Some advantages for tundra sites are described briefly.

2 Introduction

The original methodological papers on the decay of cotton fabrics as an analogue for cellulose decomposition describe either laboratory studies (Schmidt & Ruschmeyer 1958) or field studies (Latter & Howson 1977), but without many details on the practical difficulties which may confront experimenters. This paper documents some of the problems of using this assay in cold soils.

3 Problems experienced in polar and tundra sites

This assay was first used in polar sites during the International Biological Programme (Heal *et al.* 1974), at Kevo, Finland, in July 1969; Signy Island, Antarctica, in January 1970; Hardangervidda, Norway, in June 1971; and sites in Alaska, the USSR, northern Sweden and arctic Canada in the summer of 1972. There had been trials of the method of inserting and retrieving strips at Moor House, a cold temperate site in the northern Pennines, UK, as early as 1963, but tensile strength measurements were not carried out on strips from this site until 1968.

A problem encountered in cold temperate sites, and experienced much more acutely in polar and tundra sites, is frozen ground. Strips cannot be inserted or extracted whilst the soil profile is frozen because of difficulties in making an insertion slot, a lack of closure of the soil round the strip once inserted, and the almost certain damage to the fabric if it is pulled away from a frozen soil during removal. Sites underlain by permafrost have especially short experimental periods. It is vital in these sites to ensure that the soil has unfrozen to a level below the bottom of the strip, before attempting removal.

In many exposed tundra sites, vegetation may be scanty and soil development limited. Soils may be only 10 cm deep and often contain rocks which can easily damage the strip on insertion. One way to avoid damage is first to make an insertion slot with a sharpened steel plate (Plate 3 i) (Smith & Walton 1988). However, deep peat soils are also common in the Arctic and allow easy insertion, but they are

Plate 3. Insertion and retrieval of cotton strips in subantarctic soils:
 i. A sharpened steel plate is used to make the slot
ii. The strip is inserted using a blunt former
iii. Retrieval is by removing soil block in front of strip
(Photographs M J Smith)

normally anaerobic in the lower layers, a condition disturbed by oxygen ingress during strip insertion – a point to consider in all waterlogged soils.

Repeated walking on delicate vegetation to insert and harvest strips can damage the soil environment by compaction, especially peat soils formed mainly from cryptogams, but the damage may also be significant in comparisons of grazed and ungrazed temperate sites. To alleviate this problem on antarctic moss banks, we have successfully used snow shoes.

In nutrient-poor sites, typical of many tundra sites, cotton normally decomposes very slowly, and strips must often be left in for long periods to approach a tensile strength loss of cotton (CTSL) near to 50%. It should be remembered that, in these situations, a proportion of the strips can suffer damage from the growth of plant roots through the fabric.

Another difficulty of some tundra sites is a wide disparity in decomposer activity at different levels in the profile. This method easily and quickly identifies zones of unusually high biological activity in the soil profile. To obtain adequate samples in the mid-range of CTSL

Plate 4. When decomposition is slow, the entire strip can be retrieved with ease and its position in the soil profile carefully examined, but, even when decay is very advanced, position of strip in profile can be seen before removal (Photograph D W H Walton)

for all levels in the soil, it may be necessary to harvest some strips after all the cotton has disappeared at the most active point in the profile (Plate 4). It is useful to photograph strips of this type *in situ*. Generally, however, strips should be harvested as close to 50% CTSL as possible to obtain valid results.

Where soils are shallow and dry, the soil surface can crack in summer. The cracks are often associated with the strips and lead to desiccation of the upper levels of the cloth, and possibly a carry-down of surface material into the crack. This problem can occur not only in tundra soils but in any soil type which does not close completely around the strip after insertion.

A tedious problem in Antarctica, especially on Signy Island, is bird interference. If any of the strip is left protruding above the soil surface, brown skuas (*Catharacta lonnbergii*) exert considerable efforts in pulling the strips out of the ground. This problem may be partly overcome either by pinning the top down under a plastic or glass rod or, more effectively, by trimming the cloth to the soil surface and running a location wire from the strip to a central point so that it can be found again for harvesting.

A problem first noted in strips from Signy Island was an increase in tensile strength after a period in the soil. This increase was attributed to some form of cementation between the cotton fibres, and is examined in more detail by French (1988).

Attempts have been made to use cotton strips for measuring cellulose breakdown in the sediments at the bottom of antarctic freshwater lakes (Ellis-Evans 1981). Insertion into soft sediment invariably results in folding of the strip, and the folds show significantly greater CTSL than the rest of the strip. Carry-down of surface sediment is a problem, which can result in similar CTSL levels being measured throughout the profile, although assessment of microbial numbers has established that microbial activity is limited to the top 4 cm. Maltby (1988) discusses other problems concerning insertion in aquatic or swampy sites.

4   *Advantages of use in polar and tundra sites*
The low rates of decomposition characteristic of many tundra sites mean that strips can be retrieved infrequently, yet still provide a detailed CTSL curve for the soil profile. Seasonal variation in the soil activity is very marked allowing clear partitioning of breakdown to a particular season (see Lawson 1988). The tundra microflora is generally much less diverse than in temperate or tropical soils, so that it is easier to isolate the most important decomposers of cotton.

In lake sites, there is virtually no seasonal temperature change in the sediments, so that the natural situation is equivalent to a controlled laboratory experiment. Thus, the interaction between CTSL and other factors can be explored at constant temperature.

5 References

Ellis-Evans, J.C. 1981. *Freshwater microbiology at Signy Island, South Orkney Islands, Antarctica.* PhD thesis, Council for National Academic Awards.

French, D.D. 1988. The problem of cementation. In: *Cotton strip assay: an index of decomposition in soils*, edited by A.F. Harrison, P.M. Latter & D.W.H. Walton, 32-33. (ITE symposium no. 24.) Grange-over-Sands: Institute of Terrestrial Ecology.

Heal, O.W., Howson, G., French, D.D. & Jeffers, J.N.R. 1974. Decomposition of cotton strips in tundra. In: *Soil organisms and decomposition in tundra*, edited by A.J. Holding, O.W. Heal, S.F. MacLean & P.W. Flanagan, 341-362. Stockholm: Tundra Biome Steering Committee.

Latter, P.M. & Howson, G. 1977. The use of cotton strips to indicate cellulose decomposition in the field. *Pedobiologia*, **17,** 145-155.

Lawson, G. J. 1988. Using the cotton strip assay to assess organic matter decomposition patterns in the mires of South Georgia. In: *Cotton strip assay: an index of decomposition in soils*, edited by A.F. Harrison, P.M. Latter & D.W.H. Walton, 134-138. (ITE symposium no. 24.) Grange-over-Sands: Institute of Terrestrial Ecology.

Maltby, E. 1988. Use of cotton strip assay in wetland and upland environments — an international perspective. In: *Cotton strip assay: an index of decomposition in soils*, edited by A.F. Harrison, P.M. Latter & D.W.H. Walton, 140-154. (ITE symposium no. 24.) Grange-over-Sands: Institute of Terrestrial Ecology.

Schmidt, E.L. & Ruschmeyer, O.R. 1958. Cellulose decomposition in soil burial beds. I. Soil properties in relation to cellulose degradation. *Appl. Microbiol.*, **6,** 108-114.

Smith, M.J. & Walton, D.W.H. 1988. Patterns of cellulose decomposition in four subantarctic soils. *Polar Biol.* In press.

Seasonal patterns in cotton strip decomposition in soils

D D FRENCH

Institute of Terrestrial Ecology, Banchory Research Station, Banchory

1 Summary

Cotton strip assay can give a good indication of seasonal changes in soil decomposer activity, especially in temperate and sub-polar regions. Some examples are given, with comments on the interpretation of cotton strip data in relation to environmental changes. Some of the limitations of the assay, when used for this purpose, are also discussed.

2 Introduction

Cotton strip decomposition, measured by loss in tensile strength (CTSL), proceeds at rates such that, in most temperate soils, complete CTSL takes about 3 months, while in sub-polar and polar soils it may take a year or longer. Across this range of ecosystems, therefore, the rate of CTSL can be used to examine seasonal patterns of decomposition. Some common questions discussed in relation to soil decomposer activity are given below.

— At what time of year is decomposition most rapid?
— What is the difference in magnitude between the fastest and slowest seasonal rates?
— Do CTSL profiles (pattern of change with depth in the soil) vary during the year?

Three separate studies, in sub-antarctic (South Georgia), moorland (Glen Dye, Kincardineshire), and forest ecosystems (Gisburn, Lancashire), are used to illustrate the use of cotton strip assay to address these questions.

3 Sub-antarctic

Differences between CTSL rates and patterns of change with depth in soil profiles within sites on South Georgia were largely related to soil temperature and moisture changes, especially winter freezing, summer drought (in surface layers) and waterlogging (with consequent anoxia) at depth in mires (D W H Walton pers. comm.). These relationships may either be direct effects of soil microclimate on microbial activity or due to other causes, such as changes in rhizosphere exudates, soil microflora, etc.

4 Moorland

Freezing, drought and waterlogging are 3 factors which are said frequently to control soil organic matter decomposition rates. At first sight, this appears to be true in a comparison of changes in CTSL down soil profiles in a Scottish heather (*Calluna vulgaris*) moor. The cotton strips were inserted by the horizon method of French and Howson (1982) during 3 periods in the year (Table 1). The general increase in maximum decay rates is approximately related to temperature sum (deg–days >0°C). The trends down the soil profiles

Table 1. Cotton strip tensile strength loss (CTSL) at Glen Dye, during 3 contrasting periods in 1979–80, compared with some summary climatic data[1]

Period	Strip position[2]	CTSL Mean ± SE[3]	Deg-days (>0°C)	Precipitation (mm day^{-1})	Days with precipitation
October–March	Top	36 ± 1.4	469	3.03	59
	Side	36 ± 2.4			
	Bottom	25 ± 1.9			
	Whole strip	32 ± 1.3			
March–June	Top	36 ± 2.1	925	2.07	37[4]
	Side	46 ± 2.7			
	Bottom	48 ± 1.2			
	Whole strip	43 ± 1.3			
June–August	Top	44 ± 1.2	945	2.96	62[5]
	Side	46 ± 1.8			
	Bottom	42 ± 1.9			
	Whole strip	44 ± 1.0			

[1] Climatic data are from the nearest meteorological station (Banchory)
[2] Approximate depths are top = <1 cm, side = 1–5 cm, bottom = >5 cm
[3] n = 20 top and bottom, n = 10 side
[4] Nearly all rain fell in first 2 weeks or last week of this period
[5] Very even distribution of rainfall over the period

also indicate inhibition (i) by waterlogging (when not frozen) in bottom segments of cotton strips over winter, and (ii) by drought in surface segments of cotton strips in late spring and early summer, when nearly all the rain fell mostly in early April. No obvious soil microclimatic constraints appeared to operate in July and August, which were warm throughout and when rainfall was more evenly spread in time. Differences in CTSL between depths were significant (ANOVA, $P<0.05$) in the first 2 periods, but not in the third.

If, however, nutrients or carbohydrates were added to the soil before inserting the cotton strips (French 1988a), the CTSL differences between soil layers were nearly all eliminated or, in a few cases, very much reduced (Figure 1). Especially large increases in CTSL occurred at the depths where decomposition was apparently most limited by soil climate. Clearly soil climate variation does not explain entirely the patterns of CTSL with soil depth; these aspects are discussed further elsewhere (French 1988b).

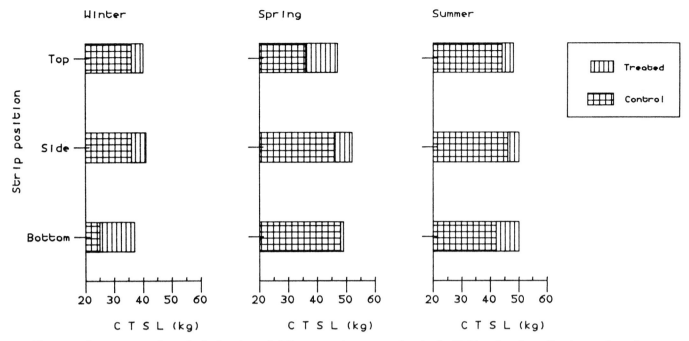

Figure 1. Some examples of elimination of differences between depths in CTSL, after 3 applications of nutrients or carbohydrates (CHO) to the soil (see French 1988a for details of method). Differences between depths in control plots were significant (ANOVA, $P<0.05$) in winter (CHO 4 gm^{-2}) and spring (P 24 gm^{-2}) but not in summer (N 40 gm^{-2}), when the range (42–46) is, however, still greater than the range in N-treated soil (48–50). There are no significant differences between depths in treated plots

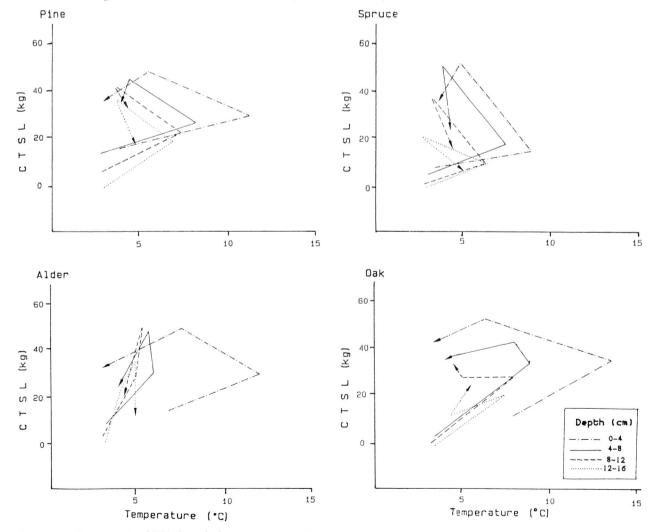

Figure 2. Changes in CTSL in relation to seasonal temperatures (measured by sucrose inversion) at 4 depths under 4 tree species at Gisburn Forest. Brown and Howson (1988) give details of method

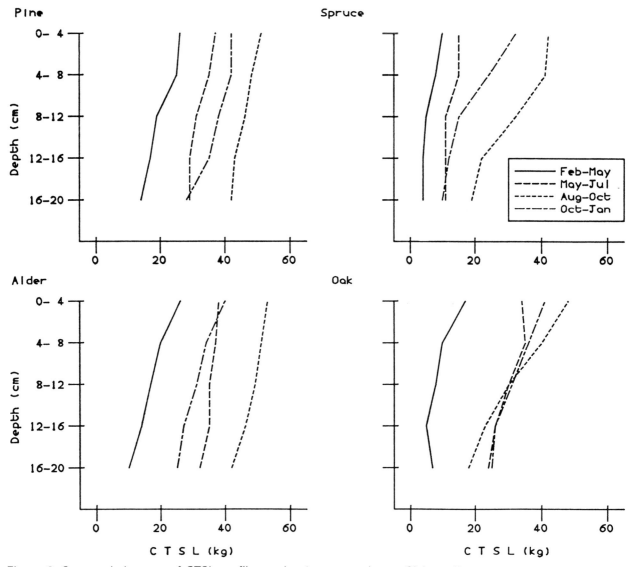

Figure 3. *Seasonal changes of CTSL profiles under 4 tree species at Gisburn Forest*

5 Forest

Cotton strips inserted in soils under 4 different tree species at Gisburn Forest (Brown 1988; Brown & Howson 1988) again showed very clear changes between seasons, in overall CTSL rates and, in some cases, in CTSL profiles (Figure 3). Somewhat similar patterns in cellulose decomposition were shown for oak (*Quercus* spp.) forest in Switzerland by Richard (1945), using cellulose cords. The CTSL data for Gisburn showed no relationship between soil moisture and CTSL at 0–4 cm, probably because soil moisture was never unfavourable. The optimal moisture content range in these or similar soil types is about 100–300% oven dried weight (OD) (Figure 12 in Heal *et al.* 1974; Figure 3c in Flanagan & Veum 1974). Similarly, any relationship of CTSL with temperature is not a simple linear response, but involves a delayed reaction (Figure 2), where changes in CTSL with temperature follow a 'cyclic' pattern, deviating markedly from the simple diagonal oscillation expected from an immediate direct relationship. As well as this lack of simple relationships between CTSL and soil climatic factors at any one soil depth, there are differences between tree species in the way CTSL profiles change (Figure 3), and these differences are not related in any clear manner to climatic patterns. Possible alternative or interactive mechanisms are discussed in detail by Brown and Howson (1988).

6 Discussion

The 3 studies referred to above indicate the value of the cotton strip assay in demonstrating seasonal patterns in decomposition processes. In each case, answers were obtained to the questions posed at the beginning of this paper. Furthermore, the observed deviations from simple relationships between CTSL and soil climatic variation forced a more detailed examination of the mechanisms of decomposition, and their implications for ecosystem functioning. In this context, it is more informative to have estimates of the full seasonal pattern of variation in decomposition rates than only a single integrated annual measurement, though that, in turn, may be better than a value for some shorter period (eg days or even hours) that cannot be related to the whole year. The cotton strip assay seems better suited to such investigations than either short-period measures such as soil respiration (with the attendant problem of root respiration) or litter weight loss which, in temperate and sub-polar regions, is generally more appropriate for longer-term estimates over several years.

However, if cotton strip assay is to be used in this way, we also need to know the extent to which it represents a general index of decomposition activity. Climatic factors seem to have similar effects on decomposition of litters (Kärenlampi 1971) and of standard cellulose substrates (Rosswall 1974), respectively, so broad agreement in seasonal patterns is likely.

I have discussed elsewhere (French 1988a) the limits to any general correlation between CTSL and litter weight losses over a complete annual cycle under different edaphic conditions. I concluded that CTSL can be taken as a reasonable general index for comparisons over a wide environmental range but, for finer analyses, the agreement between cotton strip assay results and litter weight loss, or between weight loss of different litters, is not sufficiently close for any one substrate to be used as a surrogate for any other(s). Is this also true of seasonal patterns?

Unfortunately, there appear to be very few studies which include both cotton strip assay and other measures of decomposition in directly comparable seasonal samples. In 4 sites in Norway, both litter bags and cotton strips were sampled on 2 occasions during a year, from comparable depths in the soil. In 2 sites, there was an equivalent pattern (summer/winter) in both cotton strips and plant litters; in one site some litters behaved like cotton strips and some (particularly bryophyte litter) did not; while in the fourth site no discernible seasonal pattern in CTSL was detected but litter weight losses increased in summer (cf data quoted in Heal & French 1974; Heal et al. 1974, from Hardangervidda sites). These very limited data indicate a fair degree of agreement among cotton strip and litter bag measures, but with enough exceptions to make any detailed relationship somewhat doubtful. As with annual rates, some caution is needed in the interpretation of seasonal patterns in CTSL as an indication of the patterns expected in decay of any other substrates, 'natural' or otherwise.

Finally, a warning must be given of the dangers of extrapolating annual estimates from a single period within the year. In both the moorland and the forest studies, the data from any single assay period, even if linearized and corrected for the 'average' expected effects of overall macroclimate, as in Hill et al. (1985), would give a highly biased estimate of annual loss rates or patterns of decomposition down soil profiles. For studies comparing different site or management effects, it is essential to have at least one full year's data made up of several seasonal measurements in order to estimate seasonal patterns in decomposition rates.

7 References

Brown, A.H.F. 1988. Discrimination between the effects on soils of 4 tree species in pure and mixed stands using cotton strip assay. In: *Cotton strip assay: an index of decomposition in soils*, edited by A.F.Harrison, P.M. Latter & D.W.H. Walton, 80-85. (ITE symposium no. 24.) Grange-over-Sands: Institute of Terrestrial Ecology.

Brown A.H.F. & Howson, G. 1988. Changes in tensile strength loss of cotton strips with season and soil depth under 4 tree species. In: *Cotton strip assay: an index of decomposition in soils*, edited by A.F. Harrison, P.M.Latter & D.W.H. Walton, 86-89. (ITE symposium no. 24.) Grange-over-Sands: Institute of Terrestrial Ecology.

Flanagan, P.W. & Veum, A.K. 1974. Relationships between respiration, weight loss, temperature and moisture in organic residues in tundra. In: *Soil organisms and decomposition in tundra*, edited by A.J. Holding, O.W. Heal, S.F. MacLean & P.W. Flanagan, 249-278. Stockholm: Tundra Biome Steering Committee.

French, D.D. 1988a. Patterns of decomposition assessed by the use of litter bags and cotton strip assay on fertilized and unfertilized heather moor in Scotland. In: *Cotton strip assay: an index of decomposition in soils*, edited by A.F. Harrison, P. M. Latter & D.W.H. Walton, 100-108. (ITE symposium no. 24.) Grange-over-Sands: Institute of Terrestrial Ecology.

French, D.D. 1988b. Some effects of changing soil chemistry on decomposition of plant litters and cellulose on a Scottish moor. *Oecologia.* In press.

French, D.D. & Howson, G. 1982. Cellulose decay rates measured by a modified cotton strip method. *Soil Biol. Biochem.,* **14,** 311-312.

Heal, O.W. & French, D.D. 1974. Decomposition of organic matter in tundra. In: *Soil organisms and decomposition in tundra*, edited by A.J. Holding, O.W. Heal, S.F. MacLean & P.W. Flanagan, 279-310. Stockholm: Tundra Biome Steering Committee.

Heal, O.W., Howson, G., French, D.D. & Jeffers, J.N.R. 1974. Decomposition of cotton strips in tundra. In: *Soil organisms and decomposition in tundra*, edited by A.J. Holding, O.W. Heal, S.F. MacLean & P.W. Flanagan, 341-362. Stockholm: Tundra Biome Steering Committee.

Hill, M.O., Latter, P.M. & Bancroft, G. 1985. A standard curve for inter-site comparison of cellulose degradation using the cotton strip method. *Can. J. Soil Sci.,* **65,** 609-619.

Kärenlampi, L. 1971. Weight loss of leaf litter on forest soil surface in relation to weather at Kevo Station, Finnish Lapland. *Rep. Kevo Subarct. Res. Stn,* **8,** 101-103.

Richard, F. 1945. The biological decomposition of cellulose and protein test cords in soils under forest and grass associations. I. The method of determining biological soil acidity by the so-called 'tearing' test. *Mitt. schweiz. Anst. forstl. VersWes.,* **24,** 297-397.

Rosswall, T. 1974. Cellulose decomposition studies on the tundra. In: *Soil organisms and decomposition in tundra*, edited by A.J. Holding, O.W. Heal, S.F. MacLean & P.W. Flanagan, 325-340. Stockholm: Tundra Biome Steering Committee.

ASPECTS RELATED TO USE OF THE ASSAY AND SOME EXPERIMENTAL APPLICATIONS

The colonization and decay of cotton by fungi in soil burial tests used in the textile industry

A R M BARR

Catomance Limited, Welwyn Garden City

1 Summary

The colonization and decay of cellulose by microfungi was examined by investigating the microbial populations of various soil samples, using a screened substrate technique. Cotton cloth was chosen as the substrate because of its high degree of uniformity and its suitability for tensile strength testing.

The soil burial testing of materials is discussed, as also are the aggressiveness and stability of the populations of cellulose-destroying microfungi present in the soil samples used. The necessity for standardization of test conditions is emphasized.

2 Introduction and historical review

In view of the confusion, uncertainty and often the resulting frustration that impinges on those immediately concerned with decisions based on the results of biological tests, it was felt that an attempt should be made to present a realistic picture of the situation surrounding the mystique of the soil used in soil burial tests.

In the microbial population of soil, the saprophytic fungi play a major role in converting complex materials of plant and animal origins into simpler compounds, which can be re-utilized by green plants. The decomposition part of this cycle is referred to as biodeterioration, when it involves materials of commercial value. Microbial biodeterioration is basically, therefore, the study of the behaviour of saprophytic microorganisms.

Soil is the natural habitat of such organisms, and it is logical to utilize it in any study designed to investigate the susceptibility of a substrate to biodeterioration. The simplest and certainly the easiest way of checking on the susceptibility of a material to biodeterioration is to bury it in, or place it in contact with, soil, and then to measure its response with time to the act of burial. Measurements on the effects of burial are usually assessed by calculating the loss in mass or the reduction in strength of the substrate. Thus, any test utilizing soil provides a quick way of determining the resistance of a substrate to a wide variety of different fungi.

When this kind of test is conducted out-of-doors, it is often referred to as a field test, as is the case with the well-known stake tests used for the evaluation of wood preservatives. However, under field test conditions, it is only possible to standardize the test specimens and the way in which they are buried, because the weather conditions during the test period will give rise to variations in the moisture content of the soil, its temperature and packing density. This latter parameter can have a significant influence on the aeration of the soil, and thus give rise to a change in the microbial activity within the soil. However, these limitations in no way negate the value of outdoor burial tests. In fact, the results of such tests have made a significant contribution to our knowledge of material deterioration. The major objection to the field testing of preservative-treated material is frequently one of economics, in that test sites are often costly to maintain and the tests themselves usually take longer to perform than the more compact laboratory procedures. On the other hand, this disadvantage is counterbalanced by the fact that the results of field tests generally correlate more closely with end-use situations than do the results of accelerated laboratory tests. It is, nevertheless, vital that there is available a dependable accelerated method for testing the serviceability of materials. In the last 50 years, a great deal of time and effort on the part of a large number of research workers has gone into trying to achieve such a satisfactory soil contact laboratory test method.

The primary requirement of any test method is that it should give some indication of how the material under test is likely to behave in practice. The more precisely it can do this, the better the method, and, as already indicated, a further requirement imposed by commercial considerations is that the processes of decay should be speeded up, so that a judgement on a material's resistance to biodeterioration can be obtained in the shortest possible time. This requirement applies particularly to the testing of textiles. It is interesting to note that, of the many novel soil contact tests devised for testing textiles, few, if any, of them have achieved official recognition, other than those based on straightforward burial procedures. Many of these test methods have been described and critically reviewed by Siu (1951).

Many species of micro-organisms are actually or potentially present in any soil. Their survival and activity is related to soil conditions. It is, therefore, necessary

Table 1. Test conditions specified for 6 official soil burial test methods

Country and method	Temperature (°C)	Soil pH	Moisture content (% oven dry weight)	Incubation conditions
USA	28±1	5.5–7.0	25–30	Temperature only (covered)
France	30±1	6.0–7.5	25	Temperature/RH (open)
India	30±2	—	22–25	Temperature/RH (open)
South Africa	30±2	—	25–30	Temperature/RH (covered)
Sweden	30±2	6.5–8.0	20–30	Temperature/RH (open)
Canada	29±1	—	25	Temperature/RH (open)

RH, relative humidity

for us to determine the soil conditions in order to standardize the test. Generally, an acid soil usually of pH 5.5–7.5, with high microbial activity, and a sufficiently porous nature to permit the diffusion of air and water, has been found to be suitable for testing purposes. Methods for testing a material's resistance to attack by micro-organisms have imposed conditions on the following aspects of testing procedures:

i. the size and shape of the test specimen;
ii. the manner of burial;
iii. the size and composition of the containers used for the soil;
iv. the moisture content of the soil;
v. the pH of the soil;
vi. the temperature of the soil;
vii. the method of incubation.

Table 1 lists some of the conditions of 6 official test methods for assessing the resistance of textile materials to attack by micro-organisms under conditions of soil burial.

In any critical examination of testing procedures, it immediately becomes apparent that the so-called trivial factors, such as soil contact with the test specimens, soil particle size and packing density, receive little or no attention. Furthermore, there is seldom any reference in official specifications to the management of soil beds. It is, therefore, not surprising that there has been considerable difficulty in achieving reproducibility of results from test to test, or even within an individual test.

3 The soil, micro-organisms and substrate screening

The soil contains many different kinds of organisms. The natural consequence of this fact is that the addition of a food supply to the soil will cause the population to adjust itself, as regards numbers of individuals and balance of actively involved species, to meet the new conditions. Thus, for a time, organisms of one type or another become the controlling members of the population. This state of equilibrium continues either until the food supply runs out or until the conditions developing within the soil become unsuitable for any further activity on the part of this group of the population. A new set of conditions is produced, and then another set of organisms develops

and so on, until the food supply is eventually exhausted.

Eggins and Lloyd (1968) described a new technique for the isolation of cellulolytic fungi from soil, and gave us a tool to obtain a better understanding of the microbial ecology of biodeterioration in soil. The cellulosic substrate was chromatography paper and was protected from direct contact with soil by the use of glass fibre fabric tape. The combination of substrate and screening tape was then buried, the result of which was to select, out of the microbial population present in the soil, only those organisms actively responsible for the colonization and decay of the paper cellulose. Furthermore, this screening of the substrate from direct contact with the soil minimized the risk of cultivating contaminants from the soil during subsequent subculturing of the actual colonizers for identification purposes.

Table 2. Summary of soils used in test beds

i. Meadow soil, Clent, N Worcestershire, UK (Allsopp & Eggins 1972)

ii. John Innes potting soil no.1 (Barr 1978, Table 4)

iii. Garden soil, Welwyn Garden City, Hertfordshire, UK; sand; well-rotted horse manure in 1:1:1 mixture (Barr 1978, Table 5)

iv. Mixture as (no. iii), constituted for commercial soil burial beds used 11 years, since 1965 (Barr 1978, Table 6)

v. John Innes potting soil no.1, used 7 years, since 1969 (Barr 1978, Table 6)

vi. As no. v, used one year, since 1976 (Barr 1978, Table 6)

Allsopp and Eggins (1972) used this technique to evaluate the effects of substrate screening on the colonization and decay of cotton cloth under conditions of soil burial. Boiling tubes supporting strips of screened and unscreened cloth, sterilized at 121°C for 20 minutes in an autoclave, were subjected to soil burial under controlled conditions using soil no. i in Table 2. They assessed the rate of decay of the cotton cloth by measuring the tensile strength of strips removed at intervals from the soil over a period of 3 weeks and obtained patterns of colonization by the subsequent isolation of fungal species from the textile test strips on to glucose agar and cellulose agar. Some results of this work are reproduced in Table 3, and may be summarized as follows.

i. The rate of decay in the screened cotton strips was greater than in the unscreened strips, once colonization had taken place.

ii. The ultimate tensile strength losses in both the screened and unscreened strips were approximately the same.

iii. The fungi isolated from the screened strips were predominantly cellulolytic species.

iv. More fungi were isolated from the unscreened strips of cotton cloth than from the screened strips.

It is clear from the results of this and earlier work that, by using a screened substrate technique, it should be possible to obtain a microbial profile of soil in terms of those members of the population responsible for the colonization and decay of the cellulosic substrate. Furthermore, it is also possible to equate such a profile with information relating to the aggressiveness of the soil as an agent of decay. Microbial profiles were then determined for 2 different burial soils under identical incubation conditions, and the profiles obtained were equated against the individual soils' ability to rot cellulose.

4 Method and results

Test specimens, 10 cm long and 3.5 cm wide, were cut lengthwise from the warp direction of a standard cotton test cloth and fastened on to the outer surface of 12.5 cm lengths of 3.3 cm diameter glass tubing with autoclavable, nutrient-free, non-toxic adhesive tape. All the tubes prepared in this way had glass fibre fabric tape fastened over the strips of cotton test cloth. The screened tubes were then autoclaved at 121°C for 20 minutes, cooled and divided into 2 equal sets in preparation for burial in soil nos i and ii (Table 2). At the start of the investigation, both these soils had been sieved through a standard screen and their moisture contents set at 25%, based on the oven dried weight of soil. The 2 soils in this state were then placed to a depth of 12.5 cm in separate polythene boxes measuring 60 cm long by 37.5 cm wide by 20 cm in height, and covered with sheets of glass to prevent loss of moisture during conditioning and subsequent incubation. The soil beds were left in an incubation room at 28°C for 3 weeks, the soil in each box being turned over each day during this period.

Six screened substrate tubes for each of 5 harvesting periods were then placed in each soil bed, care being taken to pack the soil loosely around each tube. They were incubated for 14 days at 28°C, and a relative humidity (RH) of 85–90% was maintained throughout the test at the air/soil interface in each box.

At intervals of 2, 4, 7, 10 and 14 days, tubes were removed from each soil bed. The glass fibre fabric tape was removed from 2 tubes, and the underlying strips of cotton cloth for isolation work were cut into sections under sterile conditions and plated out on to cellulose agar and glucose agar (Eggins & Pugh 1962). The remaining 4 strips from each soil were cleaned in

Table 3. Fungal species isolated from cotton textile buried in the test soils. Isolation tests in soil no i refer to 9 samplings over 3-week period and in soil nos ii and iii to 4 samplings over 10-day period

Soil (see Table 2) Medium	Screened cotton tests number occasions isolated						Soil burial beds presence iv–vi
	i Cellulose agar		i Glucose agar		ii	iii	
	S	U	S	U	S	S	U
Aspergillus fumigatus			4	3		1	
A. niger							+
A. sydowi							+
Chaetomium globosum	0	1			2	2	+
Fusarium oxysporum					3	3	+
F. solani	2	1					
Gliocladium roseum						2	+
Graphium sp.					1	1	+
Humicola grisea	1	3	1	1	1		+
Paecilomyces varioti							+
Papulaspora sp.	0	1	0	1	1	2	+
Penicillium chrysogenum							+
P. funiculosum	4	3	3	7	3	2	+
Peziza ostracoderma	2	0					
Rhizopus nigricans					1	2	+
Rhizopus sp.	0	1	2	5			
Sporotrichum pruinosum					2	2	+
Sporotrichum sp.							+
Trichoderma viride	4	6	5	8	3	3	+
Zygorrhynchus moelleri			0	1			
Streptomyces sp.					1	1	+

S, screened cloth; U, unscreened cloth

70% ethyl alcohol, dried at room temperature and equilibrated in a conditioning room at 21°C and 65% RH for 24 hours prior to tensile strength testing on a Hounsfield tensometer.

The similarity in the fungal profiles (Table 3) and rates of decay (Figure 1) given by the 2 soils in this controlled experiment, and the close resemblance of these 2 profiles with that of the profile obtained for the soil from Clent in Worcestershire would suggest that it is the substrate rather than the soil which determines the pattern of colonization. Furthermore, it would appear that the rate of decay is dictated by the conditions of the test, as expected. In the experiment using meadow soil from Clent, for example, it took 18 days at 25°C to achieve what soil from Welwyn Garden City at 28°C accomplished in 10 days.

In the light of these results, it was decided to examine the rate of decay and the pattern of colonization of cotton test cloth in 3 different soil beds of known age and composition, which were in normal use for the commercial evaluation of rot-proofed textiles (soil nos iv, v and vi) in Table 2). All 3 soils had been maintained under the standard soil burial test conditions (28°C with a moisture content of 25%).

This experiment (Table 3 & Figure 1) confirmed the anticipated behaviour of cotton cellulose to colonization and decay by microfungi under controlled laboratory test conditions. It also demonstrated that, as long

as the soil samples used in the test have a high population of cellulose-destroying mycoflora and the soil conditions are highly favourable to microbial activity, then similar results can be obtained, irrespective of the age or origin of the soil.

In an effort to substantiate further the validity of the screened substrate technique as a method for demonstrating the standardization of test conditions involving the use of soil, the following experiment was carried out.

A soil burial bed was chosen and its microbial profile and rate of decay were obtained. The soil was treated with a solution of a commercial biocide to deposit 1% by weight of the active chemical, technical pentachlorophenyl laurate, to British Standard 4024 (Anon 1966). The soil bed was maintained under standard soil burial test conditions of 25% moisture content based on the oven dry weight of soil and at a temperature of 28°C, with daily turning of the soil. At intervals of 5, 7, 10 and 14 days, samples of soil were removed from the soil bed and subjected to a bioassay using *Aspergillus terreus* as the assay organism. *Aspergillus terreus* was chosen because it is a common soil fungus, which is frequently found on untreated cotton cellulose and, furthermore, is known to be susceptible to the action of the biocide.

After each period of incubation, a constant-interval dilution series was prepared by transferring known

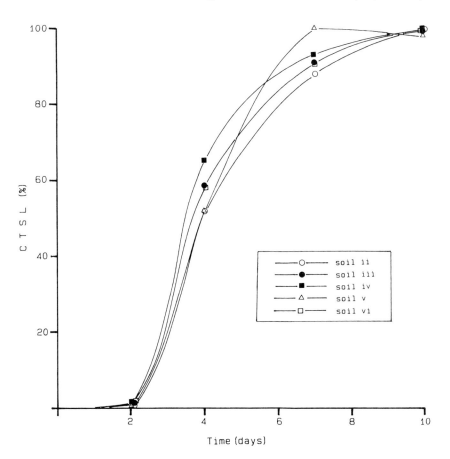

Figure 1. Tensile strength loss of screened cotton cloth during burial in soil nos ii–vi (see Table 2)

weights of the biocide-treated soil to 28 ml screw-cap Universal bottles containing 10 ml of a liquid medium. The diluted soil suspensions were then sterilized by autoclaving for 20 minutes at 121°C, cooled to room temperature and inoculated with the assay organism. Incubation was for 5 days at 28°C, after which time the fungicide/soil mixture serial dilutions were examined for the development of the assay organism.

The results of this assay demonstrated that the toxicity of pentachlorophenyl laurate in soil was diminishing with time. A screened substrate evaluation of the microbial population of the fungicide/soil mixture, 11 weeks after dosing with technical pentachlorophenyl laurate (Anon 1966), indicated that the soil microbial equilibrium was moving back towards the state which existed before the addition of the biocide.

5   Conclusions

As regards the techniques employed in textile work, they demonstrate that, under highly controlled test conditions, soil can be used to obtain reliable information on the deterioration of cellulose by microfungi.

The value of information from soil burial tests is enhanced by standardizing the test conditions of which the soil is a major part, along with the cellulose decomposing mycoflora of the soil which has been shown to be extremely stable.

6   References

Allsopp, D. & Eggins, H.O.W. 1972. A demonstration of the effects of substrate screening on the colonisation and decay of cotton textiles in soil burial tests. *Mycopath. Mycol. appl.*, **47,** 331-336.

Anon. 1966. *Pentachlorophenyl laurate.* (BS 4024.) London: British Standards Institution.

Barr, A.R.M. 1978. The role of soils in the colonisation and decay of cellulose by micro fungi. *Proc. Ann. Conv. Br. Wood Preserv. Assoc.,* 1-14.

Eggins, H.O.W. & Lloyd, A.O. 1968. Cellulolytic fungi isolated by the screened substrate method. *Experientia,* **24,** 749.

Eggins, H.O.W. & Pugh, G.J.F. 1962. Isolation of cellulose decomposing fungi from the soil. *Nature, Lond.,* **193,** 94-95.

Siu, R.G.H. 1951. *Microbial decomposition of cellulose.* New York: Reinhold.

Relationships between tensile strength and increase in metabolic activity on cotton strips

R N SMITH and J M MAW
School of Natural Sciences, Hatfield Polytechnic, Hatfield

1 Summary

Microbial activity, as measured by fluorescein diacetate (FDA) hydrolysis, has been compared with the accumulation of microbial protein and loss in tensile strength in experiments with soil burial of test fabric. The temporal patterns recorded for the 3 tests were similar. Tensile strength losses were found to occur in advance of the major increase in FDA hydrolysis and microbial protein accumulation. When cloth protected by biocides was examined, no loss in tensile strength was recorded, although a small but significant amount of FDA hydrolysis was obtained from treated cloth. The results demonstrate that FDA hydrolysis can be a reliable method for monitoring the level of microbial activity on a biodegradable material. This method is inexpensive and, unlike tensile strength measurements, permits a substrate of any shape or form to be examined for microbial activity.

2 Introduction

Studies on the microbial degradation of cellulose have been based upon measurements of tensile strength (TS), dry weight and respiration. For example, Long (1976) added powdered cellulose to soil, and found that the rise in respiration reached a peak a significant time after cotton textile buried in similar conditions had lost its tensile strength. Rubidge (1977) reported that, whereas cotton duck cloth had lost all its strength after 9 days in soil, cellulase activity in the same soil supplemented with microgranular cellulose reached its maximum after 2 weeks. The powdered or microgranular cellulose had been thoroughly mixed with the soil and should have been more available to the microflora than pieces of cotton textile. Despite this increased availability, the evidence was that microbial activity associated with cellulose degradation continued to rise and reached its maximum some time after all TS had been lost.

It is, therefore, proposed that the degradation of buried cellulose may be divided into 3 stages:

 i. initial colonization by micro-organisms;
 ii. initial degradation by enzymes released by the pioneer colonizers, resulting in tensile strength loss;
iii. proliferation of the pioneer community and perhaps other organisms, resulting in maximum enzyme activity and the complete degradation of the material.

In soil microbiology, there is a need for a method of monitoring levels of microbial activity in samples of cellulosic materials collected in the field, and in industry there is a similar requirement for monitoring the levels of microbial activity in raw materials of natural origin. Microbial degradation of cellulosic materials in natural environments such as soil has traditionally been measured by the tensile strength loss of cotton cloth (CTSL) (Walton & Allsopp 1977; Latter & Howson 1977), weight loss (Long 1976), or the evolution of carbon dioxide (Schmidt & Ruschmeyer 1958). Whilst tensile strength loss can be a reliable indicator of initial degradation, its use is limited to material which is amenable to this technique, and many natural products are thereby excluded. Weight losses take much longer to become significant and can only be used on material whose initial weight is known. Similarly, estimates of total protein can only be made if the initial protein content is known, and this technique can be time-consuming to perform. The validity of total viable counts is suspect and often bears little relation to the active biomass present on the sample (Stotzky 1972). Measurements of microbial ATP by luminescence using standard luciferin-luciferase systems are possible, but expensive, and only very small aliquots can be sampled, which would be unrepresentative of the field soil situation without considerable replication.

Microbial degradation of plant debris depends upon the production of extracellular enzymes. These enzymes are largely hydrolytic and, because in nature their substrates are heterogeneous, a battery of extracellular enzymes, eg cellulases, proteases, esterases, hemicellulases and pectinases, will be required to degrade them. A system which measures the abundance of such enzymes would give a useful indication of the extent of degradative microbial activity present. One such enzyme, esterase, is one of the degradative enzymes found in such enzyme consortia, and is present in commercial cellulase preparations. Fluorescein diacetate is hydrolyzed by esterases to release fluorescein, which can be quantified by fluorometry or spectrophotometry. Thus, the extent of esterase activity as measured by FDA hydrolysis will be proportional to the overall amount of microbial enzyme activity, which will, in turn, be proportional to the rate of enzyme degradation occurring within the sample. Schnurer and Rosswall (1982) demonstrated that FDA hydrolysis could be an effective measure of total microbial activity in soil and litter.

In the work presented below, the use of FDA hydrolysis as a measure of microbial activity on soil burial fabric has been compared with 2 established methods: measurement of CTSL and accumulation of microbial protein. The FDA hydrolysis and CTSL were then compared using untreated cloth and cloth which had been treated with 2 textile preservatives, Mystox SN and Mystox LSL. These are 2 formulations of sodium pentachlorophenyl laurate, the former being cationic and the latter anionic.

3 Materials and methods

Shirley Soil Burial Test Fabric, which complies with the requirements specified in British Standard 6085 (Anon 1981), was used as the test material in all experiments. Samples for TS determination were cut into strips 200 mm long and 35 mm wide. Cuts were made in between the dyed warp threads and then frayed to remove the first of each pair of coloured warp threads on each side of the strip, thus giving a standard number of warp threads in each sample and a frayed width of 30 mm. Strips were buried in soil using a wedge-shaped wooden stake. The centre of the strip was placed over the tapered end of the stake and pushed into the soil to a depth of 50 mm. The stake was removed and the V-shaped hole lined by the cotton strip was back-filled with soil, leaving the ends exposed.

For enzyme and nitrogen analysis, squares of soil burial fabric 25 mm x 25 mm were used and buried horizontally in the soil at a depth of 40 mm. Whilst cuts were made carefully along the weft and warp, these squares were not frayed to give standard weft and warp numbers. Both cotton strips and squares were wrapped in aluminium foil and sterilized by auto-claving. The soil used was commercial premixed John Innes no. 2 which had been passed through a 5 mm sieve, moistened to 50% water-holding capacity, and placed in square plastic bowls, 337 mm x 267 mm x 135 mm, to a depth of 75 mm. Each bowl contained approximately 10 kg of soil. Strips and squares of soil burial fabric were buried in 3 or 4 rows of 5 samples in each bowl. The bowls were then covered with 'cling-film' to prevent excessive water loss by evaporation, and incubated at 25°C for up to 7 days.

Strips for CTSL measurements were removed, shaken free of loose soil, washed gently to remove any adhering soil, and blotted on a paper towel to remove excess water. Tensile strength determinations were then made on the saturated strips using a Hounsfield Type W tensometer. For examination of enzyme activity by FDA hydrolysis or analysis of organic nitrogen, squares of soil burial fabric were removed from soil and washed gently to remove any adhering particles.

Measurements of FDA hydrolysis were performed as follows. FDA was added to acetone which had been dried with anhydrous calcium chloride to give 200 mg l^{-1} and the solution stored at –20°C. Individual squares were placed in 50 ml conical flasks containing 6 ml of pH 7.6 0.06 M sodium phosphate buffer (862.5 ml 0.06 M disodium hydrogen phosphate + 137.5 ml

0.06 M sodium dihydrogen orthophosphate); 1 ml of FDA in acetone was added and the flasks incubated for 30 minutes at 37°C in a shaking water bath. The solution was then decanted, and the amount of released fluorescein determined by measuring the absorbance at 490 nm using a spectrophotometer. The absorbance was expressed either as units per square, or units per gramme dry weight, of cloth. Samples giving absorption values greater than 0.8 were diluted before final absorbance determination. In the initial experiments, the solution was centrifuged to remove suspended material before absorbance measurements were taken. However, it was found that, after washing the strips, no suspended matter was generated and this step was omitted in subsequent experiments.

Nitrogen analyses were performed on bulked samples of 5 similar squares, using a Tecator Kjeltec semi-automatic Kjeldahl system.

In the first experiment, CTSL, accumulation of protein and increased FDA hydrolysis from samples of buried cloth were compared over 7 days. Samples were taken at one, 2, 4, 5, 6 and 7 days, and 10 strips were examined on each occasion for TS, 4 squares for FDA hydrolysis and 5 squares for nitrogen analysis.

In the second experiment, CTSL and FDA hydrolysis were compared over 7 days using untreated cloth and cloth treated with Mystox SN and Mystox LSL. Samples were taken for analysis after 2, 5 and 7 days. On each occasion, 5 strips and 5 squares were examined. Samples of treated cloth were supplied by Catomance Ltd, Welwyn Garden City, Hertfordshire.

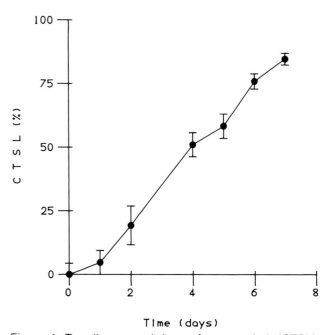

Figure 1. Tensile strength loss of cotton cloth (CTSL) during 7 days' soil burial. Vertical lines indicate 95% confidence limits

4 Results

The CTSL (%) of the cloth (Figure 1) was found to show a sigmoidal increase with time, CTSL being only 5% after one day, 50% after 4 days, 76% after 6 days and 84% after 7 days. The amount of protein (Figure 2) in the cloth showed a slight initial rise, then remained steady until day 4, and finally rose rapidly for the remaining 3 days. The increase in FDA hydrolysis with soil burial time was much more pronounced (Figure 3); the rate of hydrolysis increased gradually until day 4, then rose rapidly until day 6, with a slight increase on day 7. As with amounts of protein, the major increase occurred between days 4 and 6.

The effect of 2 textile preservatives, Mystox SN and Mystox LSL, on the CTSL and on FDA hydrolysis g^{-1} dry weight of cloth is shown in Figures 4 and 5. As in the previous experiment, the CTSL of untreated control strips was high by the seventh day (79%), whereas a slight increase in TS was recorded for the 2 treated samples. FDA hydrolysis g^{-1} dry weight of buried fabric treated with Mystox SN and Mystox LSL was only 17% and 23% of the untreated control after 7 days.

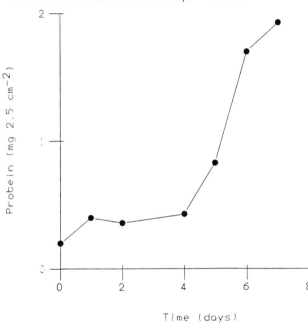

Figure 2. Increase in protein content of cotton cloth during 7 days' soil burial

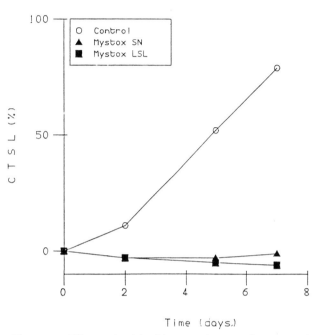

Figure 4. Effect of 2 biocides on the tensile strength loss (CTSL) of cotton cloth during 7 days' soil burial

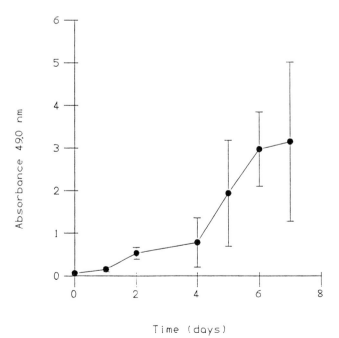

Figure 3. Increase in enzyme content of cotton cloth during 7 days' soil burial (FDA hydrolysis 2.5 cm⁻²). Absorbance due to fluorescence at 490 nm. Vertical lines indicate 95% confidence limits

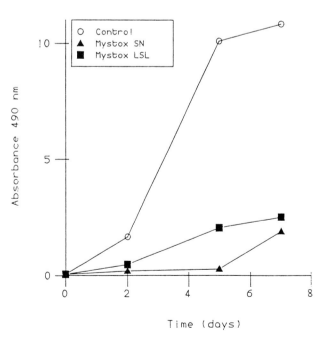

Figure 5. Effect of 2 biocides on the enzyme content of cotton cloth during 7 days' soil burial (FDA hydrolysis 2.5 cm⁻²). Absorbance due to fluorescence at 490 nm

5 Discussion

These results show that the temporal patterns of CTSL, the extent of FDA hydrolysis, and protein accumulation each followed the characteristic sigmoid curve during the 7 days of soil burial (Long 1976; Hill *et al.* 1985). When the magnitudes of these changes were compared, FDA hydrolysis gave a massive increase with absorbance, which rose from 0.06 for the control to 3.16 after 7 days, whilst protein content 2.5 cm^{-2} increased from 0.2 mg in the control to 1.928 on day 7. However, FDA hydrolysis and protein accumulation showed little increase during the first 4 days, followed by a rapid increase during the last 3 days and tailing off on day 7. Thus, the CTSL reached 50% in advance of the major increase in protein accumulation on cloth and FDA hydrolysis, and the subsequent CTSL appeared not to be affected by the increased amounts of enzymes present during the last 3 days. These results indicate that the CTSL in soil was caused by a relatively small, but active, colonizing microflora.

In contrast, complete degradation and assimilation only occur after a much longer time (Long 1976). Thus, CTSL may be considered a measure of initial stages of colonization and biodegradative attack. Similarly, Schmidt and Ruschmeyer (1958) found that carbon dioxide evolution from buried cellulose continued to rise after double the time required for a near complete TS loss. Hill *et al.* (1985) demonstrated an exponential decline in CTSL, which again resembles the sigmoid curve for CTSL reported in the present investigation and in Barr (1988). At the end of the incubation period, when biomass and enzyme levels were highest, the absolute rate of CTSL was lower than earlier in the experiment. It may be that the endo-gluconases responsible for initial attack on the long glucan polymers are only susceptible to attack at a restricted number of sites. This is a random event, resulting in an exponential decline in the number of sites attacked as the number of susceptible sites declines.

The protective action of Mystox SN and Mystox LSL has been demonstrated by both CTSL values and amounts of FDA hydrolysis. When biocide-protected cloth was examined, it was found that, even though no CTSL had been recorded, some microbial development had taken place on the protected material. The extent of FDA hydrolysis arising by biocide-treated material after 7 days was similar to untreated material after only 2 days but, whereas the untreated fabric had also suffered an 11% CTSL, no CTSL was recorded for treated cloth after 7 days. The slight increase in TS can be attributed to the lubricating effect of these treatments on material, rather than cementation as reported by French (1984, 1988). When the extent of FDA hydrolysis caused by the microflora on treated cloth was examined, a slight but significant amount of enzyme activity, in most cases between 17% and 23% of the equivalent untreated control values, was recorded. Thus, even though no CTSL had been recorded, some microbial development had taken place on the protected material.

Measurements of FDA hydrolysis and CTSL are both relatively rapid techniques whereby about 60 replicates can be recorded per hour. In contrast, analysis for total protein is a much slower procedure. The FDA hydrolysis technique has certain advantages over TS measurements. It can be performed using inexpensive reagents and standard equipment present in any biochemistry or microbiology laboratory, whilst tensile strength measurements require expensive tensometric devices. The amount of FDA hydrolyzed represents the extent of microbial activity in the whole of the sample, whereas TS measurements represent the breaking strength of the weakest cross-segment of the sample. FDA hydrolysis can also be used to determine the extent of microbial activity in decomposing materials which are in a form that renders them unsuitable for TS or weight loss determinations. It can also be employed to monitor the microbial quality of wood pulp, wool and hay, or the extent of microbial activity on decomposing leaf litter, straw or urban refuse.

6 Conclusions

Experiments with buried fabric have shown that FDA hydrolysis is a valuable indicator of microbial activity on decomposing material, and compares favourably with tensile strength loss and protein accumulation. Both hydrolytic enzyme activity and protein accumulation followed initial CTSL, implying that CTSL was caused by the relatively small pioneer microbial biomass which develops in the first stages of colonization. Tensile strength loss is only a measure of initial degradation, whereas the increase in microbial biomass and enzyme activity which will occur at all stages of degradation can be monitored by protein analysis and FDA hydrolysis. Thus, the use of TS measurements in conjunction with FDA hydrolysis can give a more complete picture of cellulose degradation.

It is proposed that, whilst a small amount of FDA hydrolysis up to 3.00 absorbance units g^{-1} may or may not indicate damage to a material, absorbance readings in excess of 9.00 units g^{-1} of material are indicative of severe microbial attack. The advantages of this technique are that the reagents are inexpensive and large representative samples can be examined rapidly to give a result within 30 minutes.

7 References

Anon. 1981. *Method of test for the determination of the resistance of textiles to microbiological deterioration.* (BS 6085.) London: British Standards Institution.

Barr, A.R.M. 1988. The colonization and decay of cotton by fungi in soil burial tests used in the textile industry. In: *Cotton strip assay: an index of decomposition in soils,* edited by A.F. Harrison, P.M. Latter & D.W.H. Walton, 50-54. (ITE symposium no. 24.) Grange-over-Sands: Institute of Terrestrial Ecology.

French, D.D. 1984. The problem of cementation when using cotton strips as a measure of cellulose decay in soils. *Int. Biodeterior.,* **20,** 169-172.

French, D.D. 1988. The problem of cementation. In: *Cotton strip assay: an index of decomposition in soils*, edited by A.F. Harrison, P.M. Latter & D.W.H. Walton, 32-33. (ITE symposium no. 24.) Grange-over-Sands: Institute of Terrestrial Ecology.

Hill, M.O., Latter, P.M. & Bancroft, G. 1985. A standard curve for inter-site comparison of cellulose degradation using the cotton strip method. *Can. J. Soil Sci.*, **65,** 609-619.

Latter, P.M. & Howson, G. 1977. The use of cotton strips to indicate cellulose decomposition in the field. *Pedobiologia*, **17,** 145-155.

Long, P.A. 1976. *The effect of industrial and agricultural biocides on the soil microflora.* PhD thesis (CNAA), Hatfield Polytechnic.

Rubidge, T. 1977. The effects of moisture content and incubation temperature upon the potential cellulase activity of John Innes no. 1 soil. *Int. Biodeterior. Bull.*, **13,** 39-44

Schmidt, E.L. & Ruschmeyer, O.R. 1958. Cellulose decomposition in soil burial beds. 1. Soil properties in relation to cellulose degradation. *Appl. Microbiol.*, **6,** 108-114.

Schnurer, J. & Rosswall, T. 1982. Fluorescein diacetate hydrolysis as a measure of total microbial activity in soil and litter. *Appl. environ. Microbiol.*, **43,** 1256-1261.

Stotzky, G. 1972. Activity, ecology and population dynamics of microorganisms in soil. *CRC Crit. Rev. Microbiol.(Chem. Rubb. Co.)*, **2,** 59-137.

Walton, D.W.H. & Allsopp, D. 1977. A new test cloth for soil burial trials and other studies on cellulose breakdown. *Int. Biodeterior. Bull.*, **13,** 112-115.

Cellulolysis of cotton by fungi in 3 upland soils

J GILLESPIE[1], P M LATTER[2] and P WIDDEN[3]
[1] *Department of Chemical and Process Engineering, University of Strathclyde, Glasgow*
[2] *Institute of Terrestrial Ecology, Merlewood Research Station, Grange-over-Sands*
[3] *Department of Biology, Concordia University, Montreal, Canada*

1 Summary

This laboratory study aimed to improve our limited knowledge of the fungal colonization of cotton strips inserted into soil for decomposer studies. The results indicated the following points.

i. Fungi cellulolytic on cotton cloth were isolated from cotton strips retrieved from the field along with many non-cellulolytic fungi possibly present as secondary colonizers.

ii. Particular cellulolytic fungi were specific to certain soils.

iii. Soil characteristics had a marked effect on the expression of cellulolysis. Some fungi shown to be cellulolytic in one soil were non-cellulolytic in other soils, although showing good growth in the latter.

iv. More cellulolytic fungi were present in a podzol soil than were isolated from the cotton strips, suggesting some selection by species. The community of cellulose decomposers in a soil appears to be adapted, or selected, to grow and decompose cellulose in the environmental conditions of that soil.

v. Pigmentation of cotton strips inserted in soils can be caused by fungi, but does not appear to be a useful indicator of particular fungi.

2 Introduction

Since the cotton strip assay was developed as an ecological test to examine cellulose decomposition rates in the field, it has generally been used without regard for the organisms involved, although data are available in the literature on the organisms colonizing cotton (Thaysen & Bunker 1927; Siu 1951; Nigam *et al.* 1960; Desai & Pandey 1971). However, cotton strips inserted into soils can be of considerable value, not only in determining the level of cellulolytic activity, but also to determine which organisms are responsible for cellulose decomposition in soil, if it can be shown how the flora developing on the cotton is related to the population of soil fungi (Widden *et al.* 1986). Degradation of the cotton in different soils is likely to be affected by different organisms, and characteristics of the soil environment may determine whether a given organism, though present, actually degrades the substrate.

The various pigments seen, when cotton strips have been retrieved from the field, were known to vary according to the soil type, vegetation or management of the site. It was of particular interest to know whether such pigments could be used as indicators of the presence of specific, and possibly cellulose-degrading, micro-organisms.

A laboratory study was therefore designed, using fungi that had been isolated from soil and from buried cotton strips in upland soils, to answer the following questions.

i. What proportion of the fungi found on cotton strips are cellulolytic on that cotton?

ii. Are species of cellulolytic fungi specific to certain soils?

iii. To what extent does the soil type affect the degree of cellulolysis generated by a given fungus?

iv. Are the strips selective for particular fungi?

v. Is the pigmentation seen on strips caused by fungi, and can it be used to indicate the presence of a particular species?

3 Method

The soils from which the non-basidiomycete fungi were isolated were located at the Moor House National Nature Reserve in the Pennines, Cumbria (brown earth under upland grassland); Gisburn Forest, Lancashire (peaty gley under Sitka spruce (*Picea sitchensis*)); and at Glen Dye, Kincardineshire (podzol under heather (*Calluna vulgaris*) moorland). They differed considerably in pH, organic matter, calcium (Ca), phosphorus (P) and potassium (K) contents (Figure 1).

Fungal communities in the Moor House site and at Banchory are described elsewhere (Widden 1987; Widden *et al.* 1986; P Widden & G Howson pers. comm.). Most of the microfungi used in this study were isolated as part of a study of the community ecology of microfungi in upland soils by one author (Widden). The fungi were isolated either from the upper 5 cm of the soil, using the soil washing method (Widden 1979), or from threads taken from the side of retrieved cotton strips as described by Widden *et al.* (1986). The general procedure is illustrated in Figure 2. Fifty soil and cotton isolates (Table 1) were chosen to include dominant species from the 3 distinctive upland soils, together with some cultures of hymenomycetes, isolated from other sites and supplied by Dr J C Frankland (Institute of Terrestrial Ecology, Merlewood Research Station).

Cellulolytic ability and pigment formation by fungi on cotton overlying each of the 3 soil types were tested using the following procedure. Soil from the upper 0–

15 cm level at each study site was separately sieved through a 5 mm mesh screen and dried to about 25% of moisture-holding capacity for storage; 14 g of soil from each site was placed separately into a 9 cm glass petri dish, and a piece of cotton cloth (Shirley Soil Burial Test Fabric, 1981 batch), 5 cm x 7 cm, was pressed on to the soil surface. Distilled water was

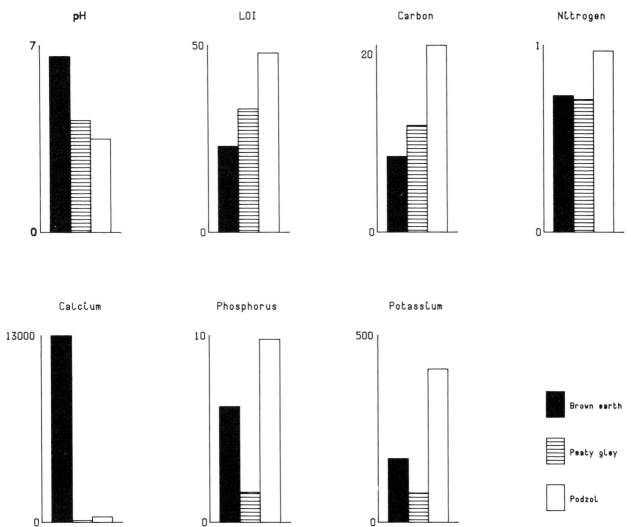

Figure 1. Characteristics of the 3 soil types used to isolate fungi and to test cellulolysis. LOI, C and N as % and Ca, P and K as µg g^{-1}

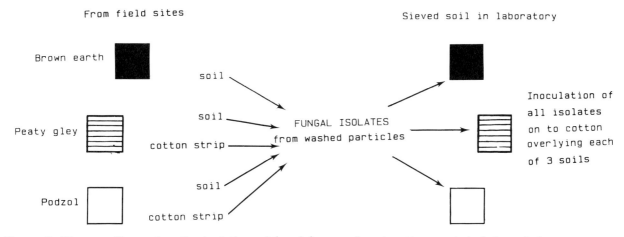

Figure 2. Diagram illustrating the isolation of fungi from soil and cotton, and their inoculation on to cotton overlying the soil

Table 1. List of fungi used in this study, with sources of isolates and occurrence on cotton. Cellulolytic ability is given for cotton strips and cotton cellulose plates (where tested), and as reported by Domsch et al. (1980), Domsch and Gams (1970) (D); Flanagan (1981) (F); or Gochenaur (1984) (G) on acid-swollen cellulose. Different isolates of the same species are indicated (i, ii)

				Cellulolytic ability			
	Source of isolates	Occurrence on cotton	Cotton strips	Cotton plates	Reported by D	F	G
HYMENOMYCETES							
Armillaria mellea	CBS culture	os	—				
Collybia dryophila	DW litter	os	+				
C. peronata	DW litter	os	+				
Flammulina velutipes	DW rotten wood	os	+		—		
Laccaria amethystea	DW litter	os	+		—		
Marasmius androsaceus	SS litter	os	—				
Mycena epipterygia	DW litter	os	—				
Nolanea staurospora	Moorland litter	os	—				
ASCOMYCETES							
Allescheria sp.[1]	UG soil		+	+			
Pseudeurotium sp.	SS soil	*	+				
PHYTOMYCETES							
Mucor circinelloides	UG soil	—	—	—			
FUNGI IMPERFECTI							
Ceuthospora lauri i & ii	SS cotton	**	—				
Chloridium chlamydosporis	SS soil	—	+		+		
Chrysosporium merdarium	UG soil	—	+				
C. pannorum	SS soil	*	+		+	+	
Chrysosporium sp.	UG soil	—	+	+			
Cylindrocarpon obtusisporum i	SS soil	*	+				
ii	UG soil		+				
Gilmaniella humicola	SS soil	*	+		+		
Humicola fusco-atra[1]	SS soil	*	+		+		
Oidiodendron tenuissimum i	UG soil	*	+				
ii	SS soil		—				
Paecilomyces carneus	UG soil	—	—		+		
Penicillium daleae i	SS soil	**	+	—			+
ii	HM cotton		+	—			
P. digitatum	SS cotton	*	—	—	+		
P. glabrum i	SS cotton	**	—	—		—	—
ii	UG soil		—				
P. janthinellum	SS cotton	**	—		+		+
P. lividum	UG soil	*	—	—	+		
P. melinii	SS soil	**	—				—
P. montanense	HM soil	—	—				
P. spinulosum i	SS soil	*	—	—	+		+
ii	HM cotton		—	—			
P. thomii	SS soil	*	+		+	+	—
Penicillium spp.	UG soil	—	—	—			
Phialophora sp.	SS soil	*	—				
Thysanophora penicillioides	SS soil	**	—				
Tolypocladium cylindrosporum	UG soil	—	—				
T. niveum	UG cotton	—	—	—			
Trichocladium opacum	SS soil	*	+	+/—	+		
Trichoderma polysporum	SS soil	*	+	—	+		
T. viride aggr. i	SS soil	*	+	+/—	+		
ii	UG soil		+	+/—			
STERILE MYCELIA i	SS soil	—	—	—			
ii	SS soil		—	—			

[1] Unconfirmed identification
** indicates major colonizers, ie 10% of isolates on one or more cotton strips from either Sitka spruce or heather moorland sites
SS, Sitka spruce; UG, upland grassland; HM, heather moor sites; DW, deciduous woodland; os, isolated in other woodland studies; CBS, Centraalbureau voor Schimmelcultures, Baarn

added to each dish to bring the moisture close to field capacity, and the dishes were sterilized by autoclaving. Each fungus was then inoculated separately on to the surface of the cotton, using an inoculum cut from the actively growing edge of an agar plate culture. The soil plates were incubated at room temperature for 9 weeks. Three replicates were prepared for each fungus and each soil.

The fungal growth on the strips was photographed at 4 and 9 weeks and the linear spread of mycelium was recorded after 9 weeks. After incubation, the cotton was removed and washed, first in 70% alcohol for one minute (a safety procedure to limit possible fungal infection during subsequent handling of cloth) and then with a jet of water. The strips were dried at room temperature overnight and oven dried at 50°C for 4 hours. The dried strips were then photographed, frayed to 3 cm width, and tested for tensile strength (TS) on a Monsanto Type W tensometer using a jaw distance of 3.5 cm.

Podzol   Peaty gley   Brown earth

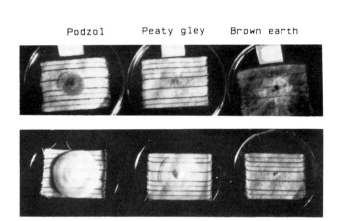

Plate 5. Growth variation of fungi on cotton overlying 3 soils
  i. Trichocladium opacum
  ii. Phialophora *spp.*
  iii. Sterile mycelium
(Photographs J Gillespie)

4 Results

Fungal growth on cotton overlying the 3 soils varied markedly with soil type, being of different growth form, extent or colour. For example, *Trichocladium opacum* and a sterile mycelium grew best and produced a black coloration of cotton on the brown earth, but formed more restricted white or grey colonies on the other 2 soils (Plate 5), whereas a *Phialophora* species grew as thick mycelium on the podzol, but sparsely on the brown earth soil. Most of the fungi

(74%) grew on all 3 soils, but some grew only on one or 2 soils. More fungi grew best on the brown earth soil (Table 2, groups 1a, b, c) than on the other 2 soils (groups 2 and 3). The brown earth was the most fertile soil in terms of higher pH and calcium, and lower organic matter (Figure 1). Other fungi showed no particular preference for any one soil, but grew equally well on 2 or all of the soils (Table 2, groups 4a–c); these fungi include many of the *Penicillium* spp. There was a tendency for fungal species to show best growth on the soil from which they were isolated.

For any one fungus, the same one of the 3 soils usually showed highest rate of cellulolysis and the most growth, but the degree of cellulolysis, as measured by tensile strength loss of cotton (CTSL), was more dependent on the soil type on which the fungus was incubated (Figure 3). Thus, many fungi which grew well on a particular soil produced no CTSL on that soil. Only one fungus, *Trichoderma polysporum,* was cellulolytic on all 3 soils. As with fungal growth, the highest CTSL values occurred on the brown earth soil among a group of fungi which, although they grew, produced no significant CTSL on the other 2 soils. Another group of fungi produced their highest CTSL on the peaty gley soil, while only 2 fungi, *Laccaria amethystea* and *Oidiodendron tenuissimum,* favoured the podzol (Figure 3). Those fungi producing the highest CTSL on either the peaty gley or podzol soils also produced some CTSL on other soils, and were from the group (Table 2, group 4) which showed no particular growth preference for any soil.

Generally, a fungus gave the highest cellulolytic activity on the soil from which it had originally been isolated (Figure 3). Two fungi, *Penicillium daleae* and *Trichoderma viride,* each had 2 cellulolytic isolates isolated from different soils, but both isolates of each species gave the highest CTSL on the same soil, suggesting that selection for genetically different ecotypes in the different soils of origin had not occurred. Where different species from the same genus were tested, there was also a tendency for these fungi to have the highest activity on the same soil. Thus, *Chrysosporium, Cylindrocarpon* and *Collybia* species were most active on the brown earth, whereas both species of *Trichoderma* were most active on the peaty gley soil.

Twenty-four fungi did not produce any significant CTSL, although some had been isolated originally from cotton strips (Table 1).

Pigmentation of the cotton strips in culture was produced by 23 fungi, but sometimes only on certain soils, and it was usually of indistinct greyish brown colours. *Trichocladium opacum* and *Chrysosporium pannorum* produced a reddish pigment only on the brown earth soil (Plate 8 i), whereas *C. merdarium* produced different coloration on 2 of the soils (Plate 8 ii). Colours can, therefore, indicate fungal colonization, but the colours

Table 2. Relative linear spread of fungi inoculated on to cotton placed on the 3 soils
i. Fungi which grew best on one soil listed and grouped under that soil, with initials after species names indicating where good growth also occurred on other soils

Brown earth		Peaty gley		Podzol	
1a *Allescheria* sp. *Chrysosporium* sp. *Flammulina velutipes* Mucor circinelloides Philalophora sp. Tolypocladium cylindrosporum		2a Marasmius androsaceus		3a None	
1b *Chryosporium pannorum* Sterile mycelium	P P	2b Armillaria mellea	B	3b Thysanophora penicillioides	G
1c Nolanea staurospora Paecilomyces carneus Penicillium sp. Tolypocladium niveum *Trichocladium opacum*	G,P G,P G,P G,P G,P	2c *Chloridum chlamydosporis* *Gilmaniella humicola* *Oidiodendron tenuissimum* i & ii	P P P P	3c *Collybia dryophila* *Collybia peronata* *Laccaria amethystea*	B B B
		2d Chrysosporium merdarium	B,P		

ii. Fungi which showed no preference for any single soil, but grew equally on 2 or 3 of the soils

Brown earth and peaty gley		Brown earth, peaty gley and podzol		Peaty gley and podzol	
4a *Humicola fusco-atra* *Pseudeurotium* sp. *Trichoderma viride*	 i	4b *Cylindrocarpon obtusisporum* i & ii P. digitatum P. glabrum P. lividum Penicillium sp. *Trichoderma polysporum* *T. viride*	 i & ii	4c Ceuthospora lauri i & ii P. montanense P. spinulosum *P. thomii* P. janthinellum	

* indicates fungi cellulolytic on cotton
B, brown earth; G, peaty gley; P, podzol

are mostly too indistinct and variable to be useful as indicators of colonizing species. Three fungi (*Allescheria* sp., *Humicola fusco-atra*, *Chrysosporium* sp.), which had shown pigmentation on cotton on soil plates, produced little pigmentation when tested on Perlite in place of soil, again indicating that soil/substrate constituents can influence the colours produced on cotton by fungi.

5 Discussion and conclusions
Various cellulolytic fungi isolated from 3 distinct upland soils have been shown to grow on and to decompose cotton strip material incubated *in vitro* on the same soils after autoclaving. Of the 50 fungi tested, 22 produced a significant CTSL of the cotton overlying the 3 soils. A small number of these cellulolytic fungi were recorded at high frequencies from cotton strips buried in the field (notably *Penicillium daleae, Trichoderma polysporum* and *T. viride*), whereas a number of fungi which produced non-significant CTSL (notably species of *Penicillium*) were also frequently isolated from cotton (Table 1). In the case of the podzol, 8 out of 13 cotton isolates tested gave no significant cellulolysis on cotton. These results suggest that successful colonization of the cotton by cellulose decomposers is a competitive process, and that some colonizers may well be secondary colonizers, using glucose or other derivatives released during cellulose

decomposition. In this regard, it is worth noting that *P. spinulosum* has been shown to be cellulolytic by Flanagan (1981) and Gochenaur (1984), using acid-swollen cellulose as a substrate. This fungus has been shown by Reese and Levinson (1952) to produce only endo-β-1, 4 glucanases and therefore to be unable to attack native cellulose. It is, therefore, probable that the fungus can attack the acid-swollen cellulose because the partial hydrolysis opens up the cellulose to attack by endo-glucanases.

Chrysosporium pannorum and *Trichocladium opacum* were both cellulolytic only on the brown earth soil (Figure 3), and were not recorded, or had very low frequency, on the cotton strips from the podzol, so the cellulolytic fungi which colonize cotton in a soil may, in part, be limited to those able to decompose cotton in that soil. However, *Chloridium chlamydosporis* and *Gilmaniella* sp., particularly abundant in the podzol, but not recorded on cotton strips, did cause CTSL on this soil in the laboratory tests. Of 8 cellulolytic fungi isolated from the podzol soil, 5 were recorded on cotton strips, indicating some selection of types by cotton in that soil. The fungal community in the soil and on cotton inserted in the podzol soil at the heather moor site is fully discussed in Widden et al. (1986). (Information on colonization of cotton strips in the peaty gley soil is yet to be published.) Shawky and

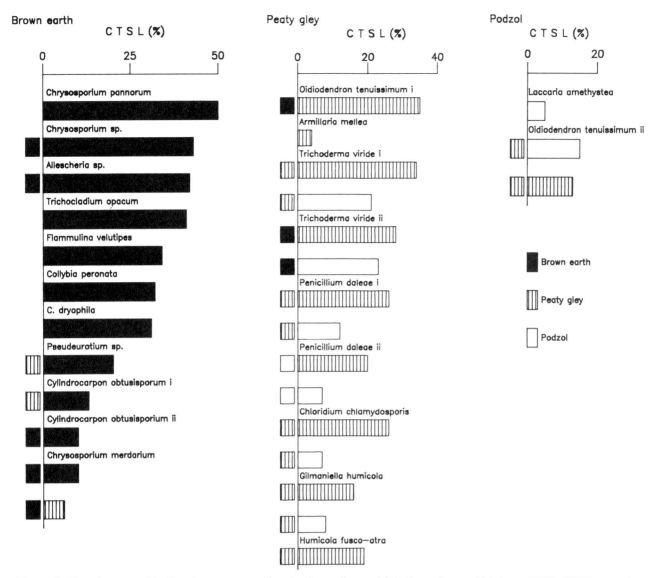

Figure 3. Fungi grouped in 3 columns according to the soil on which they showed highest CTSL. CTSL on other soils is also shown. Smaller hatched blocks indicate origin of isolates obtained from the 3 soils. Different isolates of the same species are indicated (i, ii)

Hickisch (1984) also reported a marked effect of soil type on the decomposition of cotton, and other cellulosic substrates, by *Trichoderma* species.

It is probable that the colonization of a cellulosic substrate is regulated by a selective process, based on an ability to compete for the substrate in a particular environment. In this regard, it is noteworthy that the most successful microfungal colonizers were *Trichoderma* species, which are fast-growing and known to be active soil antagonists. Further studies with a wider range of isolates would be needed to distinguish whether adaptation or selection was the controlling factor.

It must be emphasized that a number of features of this study are at variance with the natural situation. Single pure cultures were used, under laboratory conditions, with soil (not litter) as the substratum, possibly altered chemically and physically by sterilization. Pure cotton is foreign to soil and cannot wholly represent

dead vegetation as a substrate for colonization and cellulose decomposition. Depending on the nature of the vegetation type, cotton could be easier or more difficult to attack. The fungi which showed high activity in culture might not do so in the field, and those giving a negative result might be able to attack other cellulosic substrates under field conditions. The competitive element of a mixed microbial population was also eliminated by the test conditions.

The pigmentation of cotton produced by the fungi in these experiments, mostly greyish brown but also red (Plate 8), may be similar to the blackish and reddish coloration of cellulose residues from cellulose decomposition reported by Thaysen and Bunker (1927). However, as the nature of pigment production by an individual fungus varied with soil type, pigmentation cannot be used as an index of either the presence of that fungus or of cellulose decomposition by it.

During this study, a cellulose agar medium was tested

as a possible screening method for cellulolytic ability on cotton, but was not successful. It contained ground cotton passing through a 63 μ sieve, with nitrate as a nitrogen source and 0.1% glucose as an ancillary carbon source. Of 20 fungi tested over a period of 7 weeks, only a few showed any cellulolysis as evidenced by visible clearing of the plates (Table 1). Clearing did not extend outside the colony and was unaffected by added glucose. *Allescheria* and *Chrysosporium* sp. showed clearing after 1–2 weeks, but clearing by *Trichocladium opacum* and *Trichoderma viride* could only be found using a microscope to detect the disappearance of amorphous material from among a larger crystalline form of the material. The earlier loss of amorphous material is also discussed by Smith (1983) in relation to various cellulosic substrates.

Three species which gave positive results on cotton strips gave negative results on the cellulose agar. In many studies where the clearing of cellulose on agar plates has been used to detect cellulolytic activity, acid-swollen cellulose (Aaronson 1970) has been used. Comparison of such results with the cotton strip data presented here suggests that this cellulose agar method was not a satisfactory indication of an ability to attack natural cellulose. Smith (1983) suggests that the differential settlement of unequally sized particulate cellulosic substrates may reduce the availability for attack at the surface of agar plates.

Our data from cotton strips suggest that the cotton strip assay is a sensitive method for detecting the cellulolytic ability of a fungus in soil. The strong influence of soil type on cellulolytic ability (Figure 3) also shows that we should be very careful when laboratory tests are used to evaluate the cellulolytic ability of fungi and then extrapolated back to field conditions. Not only are many fungi that are capable of degrading cellulose likely to give negative results using cellulose clearing, but changes in the mineral composition, pH, or other factors in soil or culture media may markedly change the behaviour of the fungus. Clearly, in this study, some fungi were effective cellulose degraders in the fertile soil, whereas others were effective in the more nutrient-poor soils. These studies support the conclusion of Park (1976), who showed that the influence of nitrogen on the cellulolytic ability of soil microfungi varied from one species to another, some species responding better to low nitrogen levels, whereas others responded better to high levels.

In conclusion, and also in answer to the original 5 questions, the following points can be made as a result of this study.

i.   Various fungi isolated from the 3 upland soils were shown to be cellulolytic on cotton overlying the 3 soils, but many non-cellulolytic fungi also occurred on the strips, possibly as secondary colonizers.

ii. Particular cellulolytic fungi were specific to certain soils.

iii. The cellulolytic ability of individual species was greatly affected by soil type, and a group of fungi that gave the highest cellulolysis on the brown earth soil could be separated from a group which favoured the podzol and peaty gley soils for cellulolysis but showed no particular growth preference between the 3 soils. In both cases, there were fungi cellulolytic on one soil which showed no activity on other soils, despite showing good growth on them. Only *Trichoderma polysporum* was actively cellulolytic on all 3 soils.

iv. Cellulolytic fungi recorded on cotton strips from the podzol were apparently selected from a community of cellulolytic organisms present in the soil, as some cellulolytic fungi present in a particular soil were not isolated from cotton retrieved from that soil. Their absence partly reflected their inability to decompose cotton on that soil.

v.   Pigmentation on cotton strips can be caused by fungi, but is not recommended as an indicator of the presence of particular fungi.

In general, cellulolytic fungi present in a soil are able to colonize and decompose cotton strips. The native population appears to be adapted to grow and decompose in the prevailing soil conditions of the site, but the study demonstrates a clear separation between growth and cellulolysis in their relative response to soil properties.

6   Acknowledgements

Dr P Widden acknowledges funding from the Natural Sciences and Engineering Research Council of Canada and allowance of sabbatical leave from Concordia University, which made the study possible.

7   References

Aaronson, S. 1970. *Experimental microbial ecology.* New York: Academic Press.

Desai, A.J. & Pandey, S.N. 1971. Microbial degradation of cellulosic textiles. *J. scient. ind. Res.,* **30,** 598-606.

Domsch, K.H. & Gams, W. 1970. *Fungi in agricultural soils.* London: Longman.

Domsch, K.H., Gams, W. & Anderson, T.H. 1980. *Compendium of soil fungi.* New York: Academic Press.

Flanagan, P.W. 1981. Fungal taxa, physiological groups, and biomass: a comparison between ecosystems. In: *The fungal community, its organization and role in the ecosystem,* edited by D.T. Wicklow & G.C. Carroll, 569-592. New York: Marcel Dekker.

Gochenaur, S.E. 1984. Fungi of a Long Island oak-birch forest II. Population dynamics and hydrolase patterns for the soil penicillia. *Mycologia,* **76,** 218-221.

Nigam, S.S., Ranganathan, S.K., Sen Gupta, S.R., Shukla, R.K. & Tandan, R.N. 1960. Microbial degradation of cotton cellulose in soil. *J. scient. ind. Res.,* **19c,** 20-24.

Park, D. 1976. Nitrogen level and cellulose decomposition by fungi. *Int. Biodeterior. Bull.,* **12,** 85-99.

Reese, E.T. & Levinson, H.S. 1952. A comparative study of the breakdown of cellulose by microorganisms. *Physiologia Pl.,* **5,** 345-366.

Shawky, B.T. & Hickisch. B. 1984. Cellulolytic activity of *Trichoderma* sp. Strain G, grown on various cellulose substrates. *Zent.bl. Mikrobiol.,* **139,** 91-96.

Siu, R.G.H. 1951. *Microbial decomposition of cellulose.* New York: Reinhold.

Smith, M.J. 1983. Differences in the types of cellulose used in the detection of cellulolytic micro-organisms on agar media. *Revue Ecol. Biol. Sol,* **20,** 29-36.

Thaysen, A.C. & Bunker, H.J. 1927. *Microbiology of cellulose, hemicellulose, pectin and gums.* London: Oxford University Press.

Widden, P. 1979. Fungal populations from forest soils in southern Quebec. *Can. J. Bot.,* **57,** 1324-1331.

Widden, P. 1987. Fungal communities along an elevation gradient in northern England. *Mycologia,* **79,** 298-309.

Widden, P., Howson, G. & French, D.D. 1986. Use of cotton strips to relate fungal community structure to cellulose decomposition rates in the field. *Soil Biol. Biochem.,* **18,** 335-337.

Decomposition of cellulose in relation to soil properties and plant growth

P M LATTER and A F HARRISON
Institute of Terrestrial Ecology, Merlewood Research Station, Grange-over-Sands

1 Summary

Cellulose decomposition in soil is often assessed on the assumption that it relates to the rate of decomposition of native soil organic matter and, consequently, to nutrient cycling and soil fertility.

This assumption was examined, in a pot experiment, by determining the weight loss of cellulose (as filter paper), the growth of 4 species of plants, and soil chemical and physical properties, on a range of 76 soils representative of 8 major soil groups collected from various parts of the UK.

There were highly significant, but low, correlations between the rates of decomposition of cellulose and the growth of all test plant species across all soils. These correlations appeared to be linked because there were also low, but highly significant, correlations between rates of cellulose decomposition and the soil properties, to which plant growth was related. However, the regressions between (i) rates of cellulose decomposition and plant growth, and (ii) cellulose decomposition and soil properties showed significant differences in intercept for various soil types. These findings point to strong interactions of soil type in the inter-relationships of cellulose decomposition, plant growth and soil properties. The interactions with soil type are important and need to be taken into account in any interpretation of cellulose decomposition in terms of soil fertility.

2 Introduction

Decomposer processes, being an essential part of nutrient cycling, can be expected to relate to soil fertility, and Swift *et al.* (1979) stress the importance of biological (ie decomposer) processes in the replacement of available soil nutrients. It can then be inferred that, because cellulose is a major part of fresh organic matter, its rate of decomposition should relate to the rate of decomposition of soil organic matter, and any measure of cellulose decomposition, such as the cotton strip assay, would therefore serve as an index of soil fertility. This interpretation is implied in some work, with some limited support of experimental work for its general validity, but the complexity of the links in the argument is quite obvious.

Berg *et al.* (1975) considered that a standardized type of pure cellulose could act as a 'model' substance for decomposition in tundra studies, and emphasized its use to separate environmental factors from effects of litter quality. Thus, direct comparison of decay of litters and of cellulose would identify differences due to litter

quality, but also similarities related to climatic and soil factors, as shown by French (1988) when comparing weight loss of litters and cotton strip decay.

Thus, only a general similarity between relationships for decomposition of cellulose (cotton strips) and of plant litters is shown by Heal *et al.* (1974), comparing regression surfaces with site components for a range of tundra sites. Fox and Van Cleve (1983) have found a curvilinear relationship between loss in weight of filter paper and Jenny's 'k' measure of soil organic matter decomposition.

To what extent, then, can the rate of cellulose decomposition in soils provide any index of their fertility? This concept is further examined in this paper by relating directly the rate of decomposition of a single cellulose substrate (filter paper) and the growth potential of 4 plants with some chemical properties in a range of soils in a pot experiment. The results are considered relevant to the interpretation of the cotton strip assay, which was originally described only as providing a comparative index of cellulose decomposition rates in soils.

3 Method

Seventy-six UK soils (a subset of 104 soils) representing 8 different soil types (see Figure 1), with 8–10 of each type, were taken from the top 20 cm (the plant

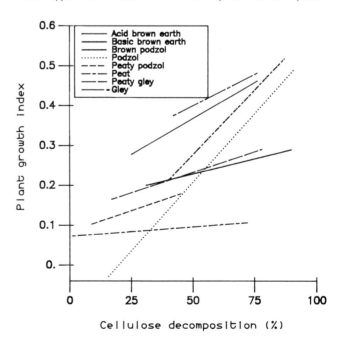

Figure 1. Interaction of soil type in the relationship of plant growth index to cellulose decomposition in 8 soil types

rooting depth) of the profiles sampled from various parts of the country (Harrison & Hornung 1983). The soils were mixed, passed, in the fresh condition, through a 19 mm mesh sieve screen (Benham & Harrison 1980), and potted, the pots being placed outside in a replicated, randomized block design on gravel beds. Cellulose decomposition rates were determined by burying 3 pieces, each 8 cm², of Whatman no. 1 filter paper in 5 mesh nylon bags (5 cm x 2 cm), in separate unplanted pots of the soils for 12 months. After this period, the percentage dry weight loss of the cellulose was determined, and was used as the measure of decomposition.

Four plant species, radish (*Raphanus sativus*), common bent-grass (*Agrostis capillaris*), white clover (*Trifolium repens*) and birch (*Betula pendula*), were grown, in separate pots, on each of the soils, radish for 6 weeks but the other species for 16 months. The productivity of each plant species was determined as total dry weight of both shoots and roots per pot. A plant growth index, which has been used as a single combined measure of relative soil fertility, was assessed as follows. For each species, an average rank value was calculated from ratios of the mean dry weight production per pot of each soil to the highest value for that species. The average rank value was then derived by calculating the mean for the 4 species, and this measure is referred to as the plant growth index. The growth of each of the 4 species was highly significantly intercorrelated over all the soils, so there was little distortion in this pattern compared to that of each individual plant species. Physical and chemical soil properties were analysed on subsamples taken after the sieving stage (see Appendix to this paper).

4 Results

4.1 The relationship between cellulose decomposition and plant growth

Cellulose decomposition rate and plant growth were positively and significantly related over all 76 soils, but the overall proportions of variation accounted for were less than 35% (Table 1). The poor, though still significant, relationship for birch is possibly explained by its slower growth rate. Using multiple regression, 50.3% of the variation in cellulose decomposition

Table 1. Relationships between plant productivity and cellulose decomposition on 76 soils of 8 soil types. The plant growth index is used as the measure of plant productivity

Species	r²	F ratio (df = 1, 74)	F ratio for significance of departure from linear regression (df = 1, 73)
Bent-grass	+0.32	35***	NS
Birch	+0.07	6*	NS
Clover	+0.31	33***	NS
Radish	+0.33	36***	NS
Plant growth index	+0.37	43***	NS

NS, not significant; *P <0.05; **P <0.01; ***P <0.001

could be accounted for by the productivity of all 4 plant species. However, covariance analysis showed significant differences in slope and intercept of the regressions for the various soil types, indicating that there were strong interactions of soil type in the relationship between the cellulose decomposition and plant growth (Table 2 & Figure 1). These interactions account for the low r² values in the relationships between the individual plant species, over all the soils (Table 1).

Table 2. Interaction of soil type in the relationship of plant growth to cellulose decomposition as shown by covariance analysis of 8 soil types

	F ratios	
Species	Slope (df = 7, 60)	Intercept (df = 7, 67)
Bent-grass	1.7 NS	3.9 **
Birch	0.7 NS	1.6 NS
Clover	2.4 *	6.4 ***
Radish	0.5 NS	2.8 *
Plant growth index	1.2 NS	4.0 ***

NS, not significant; *P<0.05; **P<0.01; ***P<0.001

4.2 The relationships between cellulose decomposition and soil properties

Cellulose decomposition rate was highly significantly related to a number of soil physical and chemical properties (Table 3), the rate being negatively related to soil organic matter content and positively related to the rest. Total soil nitrogen showed no significant relationship with cellulose decomposition over all the soils, because the relationship was confounded by a positive relationship for peat soils but a negative curvilinear relationship for the other soils. No soil property accounted for more than 40% of the variation in cellulose decomposition rate, but the fact that it related to many soil properties indicates the complexity of the soil system. Using multiple regression, it was possible to account for a total of 63.4% of the variation in cellulose decomposition by the measured soil properties. However, as with the plant relationships above, covariance analysis showed significant differences in slope and intercept of the regressions for the various soil types (Table 4 & Figure 2), indicating strong interactions of soil type in the relationships between cellulose decomposition and soil properties.

Despite the complications introduced by the soil type interaction, the soil properties which appeared to be most strongly associated with cellulose decomposition across the 76 soils were the organic matter (negative), sand, pH, total phosphorus (P), extractable potassium (K) and calcium (Ca) contents (all positive), though it has to be stated that the relative importance of the soil properties varied for the soils of different soil types. In a laboratory experiment with mainly non-organic soils and relatively high pH, Szegi *et al.* (1984) accounted for 47% of the variability in cellulose decomposition in a multiple regression analysis, includ-

Table 3. Relationships between cellulose decomposition or the plant growth index with soil properties in 76 soils of 8 soil types. Quadratic regression was used for relationships with a significant departure from linear regression . F ratio is for the linear, or quadratic (bracketted) regression with each property, df = 1, 75 (linear), 2, 73 (quadratic)

Property	Cellulose decomposition		Plant growth index	
	r^2	F ratio	r^2	F ratio
Physical				
Organic matter	−0.39	(23 ***)	−0.31	34 ***
Sand	+0.36	(20 ***)	+0.37	22 ***
Silt	+0.23	23 ***	+0.44	29 ***
Clay	+0.17	(8 ***)	+0.29	15 ***
Stones	+0.18	(8 ***)	+0.08	6 NS
Chemical				
pH	+0.36	41 ***	+0.41	52 ***
Total P	+0.20	19 ***	+0.57	49 ***
Total N	0.01	1 NS	+0.12	5 *
Total Fe	+0.19	17 ***	+0.32	34 ***
Extractable Ca	+0.23	(11 ***)	+0.40	25 ***
Extractable K	+0.38	(22 ***)	+0.58	50 ***

In Tables 3 and 4, all values expressed litre^{-1} soil, except organic matter expressed kg^{-1} soil
NS, not significant; *P <0.05; ***P <0.001

Table 4. Interaction of soil type in the relationship of cellulose decomposition, or the plant growth index, to soil properties, as shown by covariance analysis of 8 soil types

Property	Cellulose decomposition F ratio		Plant growth index F ratio	
	Slope (df = 7, 60)	Intercept (df = 7, 67)	Slope (df = 7, 60)	Intercept (df = 7, 67)
Physical				
Organic matter	0.8 NS	2.9 **	1.5 NS	3.2 *
Sand	1.4 NS	4.9 ***	1.6 NS	6.4 ***
Silt	1.3 NS	4.3 ***	1.5 NS	2.2 *
Clay	0.8 NS	6.2 ***	1.3 NS	4.7 ***
Stones	5.0 ***	6.7 ***	1.3 NS	6.7 ***
Chemical				
pH	1.2 NS	5.5 ***	3.1 *	3.8 *
Total P	1.1 NS	5.2 ***	7.5 ***	1.5 NS
Total N	0.7 NS	8.3 ***	1.7 NS	8.6 ***
Total Fe	0.8 NS	5.1 ***	0.4 NS	2.7 **
Extractable Ca	2.1 NS	6.4 ***	9.7 ***	4.9 **
Extractable K	3.1 **	5.5 ***	7.4 ***	2.9 NS

NS, not significant; *P <0.05; **P <0.01; ***P <0.001

ing available nitrogen (N) and phosphorus, magnesium (negative), pH, and clay. In a further respiration experiment with added N, P, K, and cellulose powder, N, P, K, and pH accounted for 76% of the cellulose decomposition. The poor relationship of cellulose decomposition with nitrogen recorded in our study was no doubt due to our use of analyses for total nitrogen, most of which would be unavailable in many of the organic soils used.

4.3 The relationships between plant growth and soil properties
The plant growth index was significantly related to a

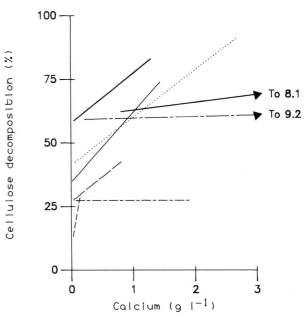

Figure 2. Interaction of soil type in the relationship of cellulose decomposition to soil calcium (extractable in ammonium acetate at pH 7.0) in 8 soil types. Key as for Figure 1

number of soil physical and chemical properties (Table 3), being related negatively to organic matter but positively to others. The proportion of variation of plant productivity which could be accounted for by each of the soil properties was generally slightly higher than for cellulose decomposition. Using multiple regression, a total of 68.6% of the variation in plant productivity could be accounted for by all the soil properties included in the analysis. However, as with cellulose decomposition above, covariance analysis showed significant differences in slope and intercept of the regressions for the various soil types (Table 4 & Figure 3), indicating that strong soil type interactions occurred in the relationships between plant pro-

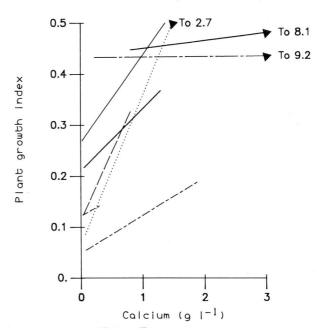

Figure 3. Interaction of soil type in the relationship of plant growth index to extractable soil calcium in 8 soil types. Key as for Figure 1

ductivity and soil properties. Despite the soil type interaction effects, the properties which appeared to be most strongly associated with plant productivity across all 76 soils (Table 3) were sand, silt, total phosphorus, extractable potassium, calcium contents and pH, but, as with cellulose decomposition, there were differences in the relative importance of properties for the various soil types.

5 Discussion

The results of this study showed that decomposition of a cellulose substrate was, in broad terms, related to the plant growth potential of 4 plants over a wide range of soil conditions. In addition, both cellulose decomposition in and plant productivity on the 76 UK soils appeared to be related, superficially at least, to similar soil properties (Figure 4). Thus, it appeared that cellulose decomposition in soils was indirectly related to plant productivity by virtue of their common dependence on the same soil properties, about 65% of the variation in each being explained by the same soil factors. Unfortunately, soil moisture, upon which both decomposition and plant growth depend heavily, was not monitored in this experiment. Variation in soil moisture conditions — water retention capacity of different soils varies in relation to their physical condition, mainly their organic matter content — may have accounted for some of the unexplained variation in decomposition.

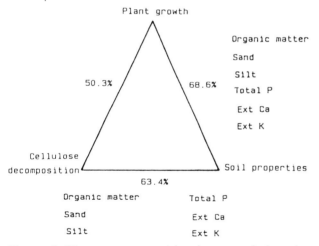

Figure 4. Diagram summarizing intercorrelations between plant growth, cellulose decomposition and soil properties in the 76 soils

There were strong interactions of soil type in the relationships of both cellulose decomposition and plant productivity to soil physical and chemical properties, and the patterns among the regressions for the 8 soil types for some soil properties were similar for both cellulose decomposition and plant growth, eg Figures 2 and 3 for extractable Ca. So, the occurrence of strong soil type interactions limits any simple interpretation of cellulose decomposition rates in terms of soil fertility across soil types, unless considerable background information is available to help in the interpretation.

6 References

Allen, S.E., Grimshaw, H.M., Parkinson, J. & Quarmby, C. 1974. *Chemical analysis of ecological materials.* Oxford: Blackwell Scientific.
Benham, D.G. & Harrison, A.F. 1980. Modification of a concrete mixer for the sieving of soils. *J. appl. Ecol.,* **17,** 203-205.
Berg, B., Karenlampi, L. & Veum, A.K. 1975. Comparisons of decomposition rates measured by means of cellulose. In: *Fennoscandian tundra ecosystems. Part 1. Plants and microorganisms,* edited by F.E. Wielgolaski, 261-267. Berlin: Springer.
French, D.D. 1988. Patterns of decomposition assessed by the use of litter bags and cotton strip assay on fertilized and unfertilized heather moor in Scotland. In: *Cotton strip assay: an index of decomposition in soils,* edited by A.F. Harrison, P.M. Latter & D.W.H. Walton, 100-108. (ITE symposium no. 24.) Grange-over-Sands: Institute of Terrestrial Ecology.
Fox, J.F. & Van Cleve, K. 1983. Relationships between cellulose decomposition, Jenny's *k,* forest-floor nitrogen, and soil temperature in Alaskan taiga forests. *Can. J. For. Res.,* **13,** 789-794.
Harrison, A.F. & Bocock, K.L. 1981. Estimation of soil bulk-density from loss-on-ignition values. *J. appl. Ecol.,* **18,** 919-927.
Harrison, A.F. & Hornung, M. 1983. Variation in the fertility of UK soils. *Annu. Rep. Inst. terr. Ecol. 1982,* 33-34.
Heal, O.W., Howson, G., French, D.D. & Jeffers, J.N.R. 1974. Decomposition of cotton strips in tundra. In: *Soil organisms and decomposition in tundra,* edited by A.J. Holding, O.W. Heal, S.F. MacLean & P.W. Flanagan, 341-362. Stockholm: Tundra Biome Steering Committee.
Swift, M.J., Heal, O.W. & Anderson, J.M. 1979. *Decomposition in terrestrial ecosystems.* Oxford: Blackwell Scientific.
Szegi, J., Gulyas, F. & Fuleky, G. 1984. Influence of soil properties on the biological activity. *Zent.bl. Mikrobiol.,* **139,** 527-536.

7 Appendix

Soil physical and chemical analyses were carried out as follows.

Organic matter as % loss in weight following ignition (LOI %) of oven dried <2 mm soil at 550°C for 2 h.

Sand, silt and clay by a differential sedimentation procedure using a Bouyoucos hydrometer (Allen *et al.* 1974); if organic matter was greater than 20%, organic matter was oxidized with boiling H_2O_2 prior to the procedure.

Bulk density (used to calculate results litre^{-1} soil) was estimated from the equation $Y = 1.558 - 0.728 (\log_{10} LOI \%)$ (Harrison & Bocock 1981), with adjustment by the value obtained for stone volume.

% stone volume as the volume of water displaced by the >2 mm fraction of the 19 mm sieved soil.

Soil pH by making a paste from fresh soil moistened to saturation point, and measuring after 30 min with a dual electrode.

Total P by nitric-perchloric-sulphuric acid digestion of air dry soil, followed by colorimetric determination by the ammonium molybdate method (Allen *et al.* 1974).

Total N by Kjeldahl digestion procedure followed by colorimetric determination by the indophenol blue method (Allen *et al.* 1974).

Total Fe by digestion in concentrated nitric-hydrofluoric acid and determination using inductively coupled plasma analysis, by N Walsh, at Kings College, London.

Extractable Ca and K by extraction in neutral ammonium acetate and determination by atomic absorption spectrometry (Allen *et al.* 1974).

Cellulolytic activity in dung pats in relation to their disappearance rate and earthworm biomass

P HOLTER

Institute of Population Biology, Copenhagen University, Copenhagen, Denmark

1 Summary
Cellulolytic activity in cattle dung pats was measured by cotton strip assay. Strips were inserted for 10 days in pats colonized by varying numbers of earthworms (Lumbricidae); measurements covered dung age 0–40 days, ie 4 10-day periods.

Cellulolytic activity was confined to dung age 10–40 days, with a maximum at 20–30 days. Activity was highest at the bottom of pats, and activity in the soil immediately below pats was higher than in the soil beyond. There was a strong, positive, correlation between rate of dung disappearance and earthworm biomass, whereas the stimulatory effect of worms on cellulolytic activity was barely significant. The data do not suggest that microbial decomposition is important to the rapid disappearance of pats, leaving physical removal of dung as the most likely explanation of the promotion of dung pat decay by earthworms.

2 Introduction
Rapid decay of cattle dung pats is characteristic of pastures with a large biomass of earthworms. In Danish pastures, 75% of pat material may disappear within only 35–45 days (Holter 1979). Further, a close positive correlation has been found between the disappearance rate of pats and the biomass of worms under and in these pats (Holter 1983). Worms seem to promote the decay of dung pats, but the actual mechanism has not been studied. The 2 most likely effects of worms in this respect would seem to be (i) simple transport, achieved by ingestion of dung and deposition of faeces elsewhere, and/or (ii) stimulation of microbial decomposition in the dung pat itself. The present paper deals only with the second possibility.

Detailed analyses of the organic components of cattle dung during decomposition are not available. Waksman *et al.* (1939) incubated horse manure in the laboratory at various temperatures. In the fresh manure, the 2 most important components were cellulose and hemicelluloses (closely followed by lignin), making up 31% and 21% (20% lignin) of the dry weight. After 47 days at 28°C, the lowest temperature used, the manure had lost 41% of its initial weight. The vast majority of this loss (96%) was attributable to disappearance of cellulose (59%) and hemicellulose (37%).

Cattle dung analysed by Lucas *et al.* (1975) contained 30% cellulose and 26% hemicellulose (based on dry weight), and so was rather similar to the horse manure. It seems reasonable to expect, therefore, that enhanced decomposition within the dung pat should manifest itself as increased cellulolytic activity. Consequently, I have tested whether cellulolytic activity in pats, assessed in field experiments as the reduction in tensile strength (TS) of cotton strips, based on the method of Latter and Howson (1977), is stimulated by the presence of earthworms. In addition, the proportion of cellulose in dung pats has been determined as a function of dung age.

3 Method
The field site is in the Strødam nature reserve, about 35 km north of Copenhagen. The experimental area was a 150 m² fenced part of a permanent pasture ('Bøgemose'), surrounded by woodland and grazed by calves and heifers. The soil is a dark, poorly drained fen soil, with a pH about 6.0 and a loss-on-ignition (LOI) of 34%. The earthworm fauna (mostly *Lumbricus festivus, L. rubellus, L. castaneus* and *Aporrectodea caliginosa*) has been described briefly by Holter (1983).

To examine cellulolytic activity in pats in relation to dung age, fresh dung from dairy cows was collected in bulk and mixed thoroughly for each of 3 experiments. The experimental pats, 22–24 cm diameter and containing 2 kg of dung, were laid out on pieces of nylon netting (7 mm mesh), which permitted easy removal of whole pats.

Each 40-day experiment included 2 batches of otherwise identical dung pats: freely exposed pats, and pats in which the invasion of earthworms was strongly delayed by one mm mesh nylon net (see Holter 1983). Each sampling comprised 3–4 pats of each type. Experimental pats were placed in the field using the following 2 types of sequence.

i. Sufficient pats for all ages in an experiment were placed in the field simultaneously, and then sampled in sequence, at 10-day intervals, from 0–30 days.

ii. Replicate samples of pat material for each age of dung were placed in the field in sequence, at 10-day intervals, so that simultaneous sampling of the pats of different ages from 10 to 40 days could all be carried out at 40 days, under the same climatic conditions. The dung was stored at –18°C until placed in the field.

The first sequence was used for 2 experiments starting on August 1983 and May 1984, and the second for the last experiment on August 1984.

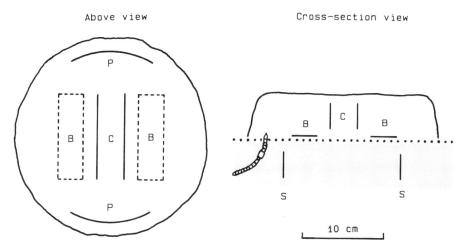

Figure 1. Positions of cotton pieces in and beneath a dung pat, viewed from above and in cross-section. Cotton positions: base (B), centre (C), periphery (P) and soil (S)

For measurements of cellulolytic activity, Shirley Soil Burial Test Fabric was used as 3 cm x 10 cm test pieces. Tensile strengths of the fabric were measured at the Danish Textile Institute, using an Alwetron F-5000 tensometer with electronic chart recorder. Test pieces were frayed down to 2 cm width and conditioned in a humidity room at 20°C and RH 65% for at least 24 hours, prior to testing in the same room.

To test the effect of within-pat position, 6 test pieces were inserted into each experimental pat, as shown in Figure 1. In the fresh dung, pieces were easily inserted while the pats were manufactured. With older pats (10, 20 or 30 days in field), the easiest procedure was to lift the pat on its net, place it upside down, remove the net, insert the pieces from below, replace the net and reposition the pat as before. This procedure involved little visible damage to the pat, as viewed from above. In 2 experiments (autumn 1983, spring 1984), pieces were also inserted into the soil under pats, with the upper edge 1–2 cm below the soil surface, and in a neighbouring area with no dung pats. Pieces put into dung (or soil) and immediately removed for testing served as field controls.

The test cotton pieces were retrieved after 10 days and the used pats removed. On the same day, the pieces were washed in cold tap water without scrubbing and dried quickly in an air current at room temperature. For practical reasons, the dry pieces usually had to be stored for 1–2 months before tensile strength testing.

Earthworms in the soil under the removed pats were forced up for collection by 0.3% formalin (Satchell 1971) from 0.125 m^2 circular areas concentric with the pats. Worms in the pats were recovered in the laboratory by hand-sorting.

Fresh weight, dry weight and organic matter content (LOI at 500°C) of the pats were determined, and subsamples of fresh dung were taken for cellulose analy-sis, using the procedure in Allen et al. (1974). The highly variable, and mostly minute, size of the dung particles created some difficulties in analysis. To overcome these difficulties, 2 fractions, ie particles >500 μm and 12–500 μm, representing practically 100% of the dung, were analysed separately as follows: 5 g of fresh dung (% dry weight and ash determined separately) was homogenized with 100 ml water in a blender to disperse particle aggregates, and passed through a 500 μm mesh sieve before washing with 5 x 100 ml water. The material remaining on the sieve was air dried at 20–25°C, weighed and processed according to Allen et al. (1974). Before delignification, this larger particulate fraction (usually 200–250 mg) comprised about 30% of the total dry weight.

For the fine-particle fraction, 10 ml aliquots were taken from the vigorously stirred fluid that had passed through the sieve. Each aliquot was filtered on a membrane filter with pore size 12 μm (Sartorius, cellulose nitrate). The moist filtrate was gently scraped off the filter, air dried and weighed. Usually, 40–50 mg dry filtrate was obtained from 5–6 aliquots. The calculated total amount of this fraction comprised 60–70% of the initial dung dry weight. The dry filtrate was then processed as in Allen et al. (1974), with sizes of flasks and amounts of chemicals scaled down by one tenth. Filtering and washing were done on 5 μm membrane filters (Sartorius, polytetrafluoroethylene with high chemical resistance), and the filtrates were air dried at room temperature.

Average temperatures in the dung (or soil, see below) during the 10-day periods were measured by the sucrose inversion method (Jones 1972; Sibbesen 1975; Jones & Court 1980), using 10 ml tubes (6.8 cm x 1.6 cm) inserted into the pats.

4 Results

Cellulolytic activity, measured as loss in tensile strength of cotton (CTSL), in relation to dung age and position within pats showed the same basic pattern

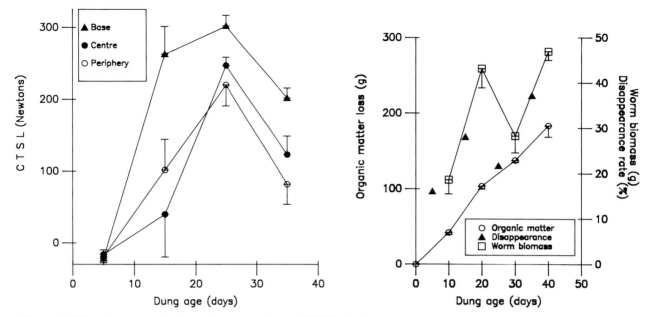

Figure 2 i.Tensile strength loss of cotton pieces (CTSL) during 4 10-day periods; freely exposed pats, autumn 1984. Points are in the middle of the periods. Field control TS 361.7 N (SE ± 5.6 N; n = 10)
ii. Mean earthworm biomass (g per pat), content of organic matter (g per pat) and disappearance rate of the same pats (% of initial dung weight) in relation to dung age
(Bars indicate standard errors)

in all experiments. However, the last experiment (autumn 1984) provided the clearest illustration (Figure 2 i), as all measurements were taken simultaneously and hence at the same temperature (second sequence). No cellulolytic activity was recorded during the first 10-day period, but a sharp rise to a maximum occurred during the third (second in the spring experiment) period, followed by a marked decline.

Before 10 days, TS values were actually about 5% higher than those of field controls in all experiments, as found in some early results for TS (Hill et al. 1985), and could be due to 'cementation', recently discussed by French (1984, 1988). Possibly, this phenomenon

was also present in strips from the other periods, but masked by the high CTSL.

The amount of dung (as organic matter) per pat in relation to age, the disappearance rate for each 10-day period (as % of dung initially present in each period), and earthworm biomass are shown in Figure 2 ii. The pats decayed fairly rapidly, the 75% disappearance time being 40–45 days, but no correlation is indicated between disappearance rates in the 4 periods and cellulolytic activity. In this experimental example, earthworms arrived early and their biomass fluctuated relatively little.

Table 1. Dung temperature, water percentage and earthworm biomass in experiments 1–3
(means of 10-day periods, with range shown in brackets)

Experiment	Pat condition	Mean temperature (°C)	Water in dung or soil (%)	Earthworms 0.125 m^{-2} (g fresh wt)	Total dung disappearance as % of initial dung
August– October 1983	F	15 (12–22)	83 (81–85)	7.5	26.4
	E	15 (12–18)	43 (36–47)	1.8	20.0
	FS & CS				
May–June 1984	F	18 (16–20)	81 (79–84)	12.9	46.9
	E	15*		3.4	42.6
	FS & CS				
August– October 1984	F	12	85 (84–86)	34.3	70.3
	E			7.7	45.6

F, freely exposed pats; E, pats enclosed to delay earthworm invasion; FS, soil below F; CS, control soil without dung
* Measurements only from dung age 10–20 days
Temperatures and water percentages from the 2 types of pats and soil were pooled as there were no between-type differences

Temperature, earthworm biomass and rate of dung disappearance varied in the 3 experiments (Table 1). In all experiments, the earthworm biomass in enclosed pats was only about 25% of that in those freely exposed.

The measurements of cellulolytic activity are summarized in Table 2. Clearly, the experiments differ as to the general level of CTSL, which was highest in spring 1984, with the highest mean temperatures. TS values (not the percentage losses given here) within each experiment were analysed by 2-way ANOVA, the 2 factors being dung age and position in pat or 'condition' (exposed, unexposed). Within condition, CTSL in the bottom of pats was always significantly ($P<0.01$ in most cases) higher than that from the centre or periphery, whereas the latter never differed significantly. The differences between condition (within positions) at $P<0.1$ are shown in Table 3. It appears that cellulolytic activity in the soil below exposed pats was significantly higher than in the control soil without dung. On the other hand, differences between the 2 conditions of dung pats were barely significant, or absent, as in spring 1984.

Table 2. Cellulolytic activity in dung pats, expressed as tensile strength loss of cotton (CTSL), compared with field controls. Values are averages of all 10-day periods

Pat/soil condition and position of pieces	CTSL (%) Aug–Oct 1983	May–June 1984	Aug–Oct 1984
FC	17.3	54.1	27.2
FP	18.5	54.0	26.7
FB	43.3	63.7	51.5
EC	8.2	56.1	25.4
EP	13.1	53.6	21.9
EB	34.0	62.7	38.3
FS	30.9	49.3*	–
CS	19.7	26.3*	–

C, centre; P, periphery; B, base
F, freely exposed pats; E, pats enclosed to delay earthworm invasion; FS, soil below F; CS, control soil without dung
* Values only from dung age 10–20 days

Table 3. Differences in tensile strength values between condition of enclosed, unenclosed, dung pats or soil, for within-dung pat positions (2-way ANOVA)

Experiment	Position	Significance
August–October 1983	C	$P<0.1$
	B	$P<0.1$
	Soil	$P<0.025$
May–June 1984	Soil	$P<0.05$
August–October 1983	B	$P<0.1$

C, centre; B, base

Figure 2 shows the lack of any close correlation between the 4 disappearance rates and the corresponding cellulolytic activities. To test whether the disappearance rate was correlated with the general level of cellulolytic activity, I calculated the following quantities, averaged over the 3 last 10-day periods: (i) CTSL % for dung pat centre + periphery (pooled because of the lack of significant differences); (ii) CTSL % at bottom of pat; (iii) earthworm biomass (g fresh weight per pat); (iv) rate of dung disappearance (disappearance of organic matter in 10 days as percentage of initial amount). The coefficients of correlation obtained were: 0.92 (i/ii), 0.07 (i/iii), 0.15 (i/iv), 0.42 (ii/iii), 0.47 (ii/iv), 0.95 (iii/iv). Significant correlations ($P<0.01$) were found only between cellulolytic activities in the 2 positions (i/ii), and between earthworm biomass and disappearance rate (iii/iv).

In Table 4, concentrations of holocellulose and α-cellulose in the 2 autumn experiments are shown. Only data from the first, third and fifth sampling are given because of the small variations within each experiment. Generally, the dung contains about 60% holocellulose and 30% α-cellulose; in the 1983 experiment, concentrations decline slightly with increasing dung age.

Table 4. Cellulose concentration (fraction of total organic matter in relation to dung age. Data from autumn experiments in 1983 and 1984

	Dung age (days)	Holocellulose	α-cellulose
1983	0	0.62	0.29
	21	0.59	0.27
	43	0.58	0.25
1984	0	0.63	0.31
	20	0.62	0.31
	40	0.63	0.29

5 Discussion

Cellulolytic ability of fungi isolated from dung has repeatedly been demonstrated (eg Fries 1955; Wicklow et al. 1980; Morinaga & Arimura 1984), and so the presence of cellulolytic activity in dung pats is no surprise. More interesting is the short duration of this activity (Figure 2). In this example, about 70% of the dung had disappeared after 40 days but, in cases when most of the dung was still present, the same decline in activity was found. This decline is not attributable to lack of natural substrate (Table 4) or to limiting nitrogen, because the C/N ratio was 12:20. Antagonism between members of the fungal population could cause the reduced activity; interference between dung fungi has been reported (Wicklow & Hirschfield 1979; Wicklow & Yocom 1981).

The present work aims to test whether earthworms promote dung disappearance by stimulating decomposition in the dung pat. It raises the following 2 questions.

i. Do worms stimulate decomposition (cellulolytic activity) in the dung?

ii. If so, is decomposition in the dung important to the rapid disappearance of pats?

76

Much work has been done on the effects of earthworms on soil micro-organisms, reviewed recently by Edwards and Lofty (1977) and Satchell (1983). Particularly relevant is the demonstration of Loquet *et al.* (1977) that worm casts and, in some cases, burrow walls contain more cellulolytic aerobes and hemicellulolytic bacteria than the intervening soil. Further, Ross and Cairns (1982) found that the presence of worms stimulated cellulase activity in some of their soil samples. Hence, earthworms may promote cellulolytic activity in the dung, eg by increased aeration and by inoculation with cellulolytic microflora in casts.

The present data do not provide conclusive evidence of such stimulation. Cellulolytic activity in the bottom layer of pats where earthworm activity is likely to be most intense is always higher than elsewhere (Table 3), but this increased activity occurred equally in experiments with practically no worms (enclosed pats, autumn 1983) and with many worms (free pats, autumn 1984). This position effect may be caused by other factors, such as aeration. In the autumn experiments, cellulolytic activity in enclosed pats tends to be lower than in freely exposed pats (with 4 times higher worm biomasses), particularly at the bottom. None of these differences are significant, however, neither are the positive correlations, from the combined data, between cellulolytic activity and earthworm biomass.

Figure 3 shows a simple model of the causal relationships leading to dung disappearance; the strength of the relations between cause and effect (disappearance rate) is given by the path coefficients (cf Sokal & Rohlf 1981). Because of the non-significant correlations between worm biomass and cellulolytic activity in centre + periphery (r = 0.07) and between the latter and the disappearance rate (r = 0.15), cellulolytic activity in centre + periphery has been excluded. The model explains 90.4% of the variation in disappearance rate, and so provides a good description. The impact from cellulolytic activity appears to be small (0.08), compared to the strong causal relationship between worm biomass and disappearance rate (0.92). In agreement with this finding, the proportion of cellulose changes only slightly with increasing dung age (Table 4).

In conclusion, it seems that, although worms may stimulate cellulolytic activity, their effect is weak and is not important in the disappearance of dung pats. Provided that microbial decomposition of the dung is accurately reflected by the cotton strip assay, we must conclude that worms do not promote dung disappearance by stimulating decomposition in the dung itself. On the other hand, the strong correlation found earlier (Holter 1983) between earthworm biomass and disappearance rate is confirmed by the present data, which leaves the physical removal of dung by worms as the most likely explanation, consistent with the increase of cellulolytic activity in the soil beneath pats. This effect may be caused by leaching of nutrients (Davies *et al.* 1962; MacDiarmid & Watkin 1972; Underhay & Dickinson 1978) from the dung into the soil. If dung is, indeed, incorporated into this soil, its decomposition is likely to be facilitated by nutrient enrichment.

6   *Acknowledgements*
I am grateful to N B Hendriksen, P M Latter and C Overgaard Nielsen for help and discussions, to H Eigtved, Dansk Textil Institut, for help with TS testing, and to R Boe Jensen, Kemisk Lab II, for loan of a polarimeter. B Brandt provided technical assistance.

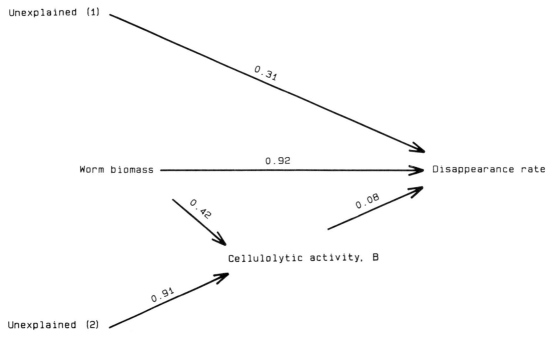

Figure 3. Path diagram relating earthworm biomass and cellulolytic activity in the bottom (B) of dung pats to the disappearance rate of these pats. Sources of unexplained variation in disappearance rate (1) and cellulolytic activity (2) respectively are indicated

7 References

Allen, S.E., Grimshaw, H.M., Parkinson, J.A. & Quarmby, C. 1974. *Chemical analysis of ecological materials.* Oxford: Blackwell Scientific.

Davies, E.B., Hogg, D.E. & Hopewell, H.G. 1962. Extent of return of nutrient elements by dairy cattle: possible leaching losses. *Trans. 4th int. Conf. Soc. Soil Sci., Lower Hutt, New Zealand,* **20,** 715-720.

Edwards, C.A. & Lofty, J.R. 1977. *Biology of earthworms.* 2nd ed. London: Chapman & Hall.

French, D.D. 1984. The problem of 'cementation' when using cotton strips as a measure of cellulose decay in soils. *Int. Biodeterior.,* **20,** 169-172.

French, D.D. 1988. The problem of cementation. In: *Cotton strip assay: an index of decomposition in soils,* edited by A.F. Harrison, P.M. Latter & D.W.H. Walton, 32-33. (ITE symposium no. 24.) Grange-over-Sands: Institute of Terrestrial Ecology.

Fries, L. 1955. Studies in the physiology of *Coprinus.* I. Growth substance, nitrogen and carbon requirements. *Svensk bot. Tidskr.,* **49,** 475-535.

Hill, M.O., Latter, P.M. & Bancroft, G. 1985. A standard curve for inter-site comparison of cellulose degradation using the cotton strip method. *Can. J. Soil Sci.,* **65,** 609-619.

Holter, P. 1979. Effects of dung beetles (*Aphodius* spp.) and earthworms on the disappearance of cattle dung. *Oikos,* **32,** 393-2402.

Holter, P. 1983. Effect of earthworms on the disappearance rate of cattle droppings. In: *Earthworm ecology from Darwin to vermiculture,* edited by J E Satchell, 45-57. London: Chapman & Hall.

Jones, R.J.A. 1972. The measurement of mean temperatures by the sucrose inversion method. *Soils Fertil., Harpenden,* **35,** 615-619.

Jones, R.J.A. & Court, M.N. 1980. The measurement of mean temperatures in plant and soil studies by the sucrose inversion methods. *Pl. Soil,* **54,** 15-31.

Latter, P.M. & Howson, G. 1977. The use of cotton strips to indicate cellulose decomposition in the field. *Pedobiologia,* **17,** 145-155.

Loquet, M., Bhatnagar, T., Bouche, M.B. & Rouelle, J. 1977. Essai d'estimation de l'influence écologique des lombriciens sur les microorganismes. *Pedobiologia,* **17,** 400-417.

Lucas, D.M., Fontenot, J.P. & Webb, K.E., Jr. 1975. Composition and digestibility of cattle fecal waste. *J. Anim. Sci.,* **41,** 1480-1486.

MacDiarmid, B.N. & Watkin, B.R. 1972. The cattle dung patch. 2. Effect of a dung patch on the chemical status of the soil, and ammonia nitrogen losses from the patch. *J. Br. Grassld Soc.,* **27,** 43-48.

Morinaga, T. & Arimura, T. 1984. Degradation of crude fiber during the fungal succession on deer dung. *Trans. mycol. Soc. Japan,* **25,** 93-100.

Ross, D.J. & Cairns, A. 1982. Effects of earthworms and ryegrass on respiratory and enzyme activities of soil. *Soil Biol. Biochem.,* **14,** 593-587.

Satchell, J.E. 1971. Earthworms. In: *Methods of study in quantitative soil ecology,* edited by J. Phillipson, 107-127. (IBP handbook no. 18.) Oxford: Blackwell Scientific.

Satchell, J.E. 1983. Earthworm microbiology. In: *Earthworm ecology from Darwin to vermiculture,* edited by J.E. Satchell, 351-364. London: Chapman & Hall.

Sibbesen, E. 1975. Temperaturmaaling med sukkermetoden. *Meddr K. Vet.- og Landbohojsk. afd. Plant. ernaer.,* no. 1101.

Sokal, R.R. & Rohlf, F.J. 1981. *Biometry.* 2nd ed. San Francisco: Freeman.

Underhay, V.H.S. & Dickinson, C.H. 1978. Water, mineral and energy fluctuations in decomposing cattle dung pats. *J. Br. Grassld Soc.,* **33,** 189-196.

Waksman, S.A., Cordon, T.C. & Hulpoi, N. 1939. Influence of temperature upon the microbiological population and decomposition processes in composts of stable manure. *Soil Sci.,* **47,** 83-114.

Wicklow, D.T. & Hirschfield, B.J. 1979. Evidence of a competitive hierarchy among coprophilous fungal populations. *Can. J. Microbiol.,* **25,** 855-858.

Wicklow, D.T. & Yocom, D.H. 1981. Fungal species numbers and decomposition of rabbit faeces. *Trans. Br. mycol. Soc.,* **76,** 29-32.

Wicklow, D.T., Detroy, R.W. & Adams, S. 1980. Differential modification of the lignin and cellulose components in wheat straw by fungal colonists of ruminant dung: ecological implications. *Mycologia,* **72,** 1065-1076.

Use of cotton cloth in microcosms to examine relationships between mycorrhizal and saprotrophic fungi

J DIGHTON and P M LATTER
Institute of Terrestrial Ecology, Merlewood Research Station, Grange-over-Sands

Poster summary

Small pieces of Shirley Soil Burial Test Fabric (cotton)(2 cm x 10 cm) were used as one of a number of substrates to investigate the effects of interactions between saprotrophic and mycorrhizal fungi in controlled culture microcosms (Figure 1). Substrates were placed inside or outside a root exclusion zone, delimited by a nylon mesh pervious to fungal hyphae but not to roots, in a series of mini-propagators used as microcosms. The complexity of organism interactions was built up in a factorial experimental design with the following levels of complexity:

   i. sterile medium (perlite:peat:nutrient solution);
  ii. fungus alone (saprotroph or mycorrhizal fungus);
 iii. non-mycorrhizal roots;
 iv. non-mycorrhizal roots plus saprotroph;
  v. mycorrhizal roots;
 vii. mycorrhizal roots plus saprotroph.

Mycorrhizal roots were shown to enhance substrate decomposition to varying degrees, but the addition of the saprotroph to the mycorrhizal system suppressed the decomposition rate. This work has been published in detail elsewhere (Dighton *et al.* 1987).

Reference

Dighton, J., Thomas, E.D. & Latter, P.M. 1987. Interactions between tree roots, mycorrhizas, a saprotrophic fungus and the decomposition of organic substrates in a microcosm. *Biol. Fertil. Soils*, **4**, 145-150.

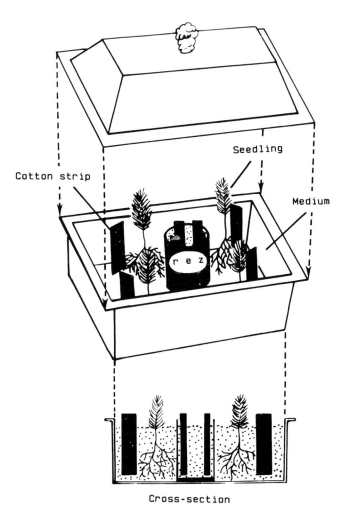

Figure 1. Illustration of microcosm used for plant–mycorrhiza–substrate decomposition interaction rez, root exclusion zone

Use of cotton strip assay to assess the effect of formaldehyde treatment on a peat soil

P M LATTER[1] and J MILES[2]
[1] Institute of Terrestrial Ecology, Merlewood Research Station, Grange-over-Sands
[2] Institute of Terrestrial Ecology, Banchory Research Station, Banchory

Poster summary
An area of 2–3 ha of low-lying blanket bog on Gruinard Island, Wester Ross, Scotland, was known to be contaminated with anthrax spores released during 1943. Formaldehyde was selected as the most suitable agent for decontamination. Prior to decontamination, the likely ecological consequences were investigated and the cotton strip assay was used among other biological tests to assess the extent and rate of recovery of decomposer activity after a sterilizing application of 5% formaldehyde solution at 50 l m^{-2} (Miles *et al.* 1988).

This example of a practical application of the assay was described. The study effectively demonstrated the early return of decomposer activity after the virtual elimination of all microbial activity, and showed that recovery was enhanced by the addition of fertilizer.

The results of the cotton strip assay gave similar results to other assessments, which were of microbial populations or activity, and all tests indicated a recovery period of about 2 months.

Reference

Miles, J., Latter, P.M., Smith, I.R. & Heal, O.W. 1988. Ecological effects of killing *Bacillus anthracis* on Gruinard Island with formaldehyde. *Reclam. Reveg. Res.* In press.

FIELD APPLICATION IN SPECIFIC ENVIRONMENTS

Temperate

Discrimination between the effects on soils of 4 tree species in pure and mixed stands using cotton strip assay

A H F BROWN
Institute of Terrestrial Ecology, Merlewood Research Station, Grange-over-Sands

1 Summary
Soil differences which are thought to have developed more than 20 years after planting under experimental stands of 4 different tree species, at Gisburn Forest, Lancashire, could not readily be detected by chemical analysis.

Cotton strip assay, however, indicated that there were clear between-species effects on soil biological activity. This method was also able to discriminate between the soils of a given tree species, according to whether or not it was grown in association with other tree species. The height growth of most of the tree species in mixture was improved, apparently through modification of soil nutrient availability, paralleled by the rate of decomposition of the cotton strips.

2 Background
It is generally agreed that soils are influenced by afforestation, and that different tree species are likely to have differing effects. Further, species in mixture may interact in their effects, both on each other and on the properties of the site. To obtain information on these points, a long-term afforestation experiment was established at Gisburn (in the Lancashire/Yorkshire border area of the west Pennines) in 1955. This experiment has enabled the effects on vegetation and soils of 4 tree species, both pure and mixed, to be monitored (Holmes & Lines 1956; Brown & Harrison 1983). The tree species planted consist of 2 conifers, Norway spruce (*Picea abies*) and Scots pine (*Pinus sylvestris*), and 2 deciduous hardwoods, oak (*Quercus petraea*) and alder (*Alnus glutinosa*). The mixed stands comprise all possible 2-species combinations. Each of the resulting 10 treatments was planted as a 0.2 ha plot in a randomized experiment with 3 replicate blocks.

The elevation of the experimental site ranged from 260 m to 290 m, with a mean annual rainfall of 1370 mm. Soils were surface water gleys, tending towards peaty gleys; the original vegetation consisted of a grazed fescue/bent-grass/mat-grass (*Festuca/Agrostis/Nardus*) sward similar to the poorest (Class 6) of Ball *et al.*'s (1981) rough pastures.

In attempting to measure changes in the soil and the effects of the different species, the question arose of which attributes should be assessed. Initially, emphasis was placed on chemistry, using standard soil analytical methods. Twenty years after planting the different stands, chemical analyses of the soils gave no consistent differences, even between the different monocultures, for most major nutrients (total nitrogen (N), extractable phosphorus (P), potassium (K) and magnesium (Mg)). Analytical methods developed for agricultural soils are not appropriate for distinguishing differences in the predominantly poor soils used for forestry. In such soils, the rate of nutrient cycling may be more important than the size of the nutrient pool (Harrison 1985), and Jacks' (1963) conclusion that the 'natural fertility of a soil is a biophysical rather that a physicochemical phenomenon' accords with this view. At Gisburn, there was evidence that, in soil biological terms, differences between the stands were developing at a relatively early stage in their life; for example, very different populations of higher fungi were evident in the different stands (Brown 1978). It was, therefore, concluded that a biological assessment of soil differences might be informative. Organic matter decomposition processes are regarded as important both in nutrient cycling and as precursors of soil chemical and pedological changes (Harrison 1985); therefore, some measure of decomposer activity was considered most relevant. Measurement of the decomposition of the different species of tree litter would be of very little value, as species differences in litter quality are confounded with the differences in site properties, developed in the different stands. A method which was integrative over a period, rather than instantaneous like most measurements of respiration, was desirable. A method that could assay the changes in properties which might have developed down the mineral soil profile, in addition to the forest floor, was also to be preferred. The cotton strip assay, providing an index of decomposer potential with a standard substrate, appeared very suitable.

3 Monoculture comparisons
Following a pilot study in the pure plots of one replicate block, a large-scale seasonal study of the 4 mono-

cultures of all 3 replicate blocks at Gisburn was undertaken. This work is described more fully elsewhere (Brown & Howson 1988), but can be summarized here. A discrete 12-week assay was repeated at 6-weekly intervals for a whole year. The pattern of tensile strength loss of cotton (CTSL) (averaged for whole strips) showed seasonal differences in decomposer activity, and between each of the 4 stands (Figure 1 in Brown & Howson 1988). Clear and significant differences between treatments were evident on 8 of the 9 occasions, and these were consistent between the 3 blocks. Although the interpretation of these differences may be open to argument (Howard 1988), good discrimination was achieved between the effects of species on soil. It demonstrated that the soils under the different trees were now different, in some way related to an important soil attribute – *viz* the potential for decomposition of organic matter. The relevance of these differences to other site parameters or processes is partly explored in the next paper (Brown & Howson 1988). Emphasis in the present paper is, however, placed on the effects of the mixed-species plots at Gisburn, perhaps the most important aspect of this long-term forestry experiment.

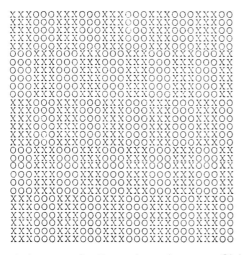

Figure 1. Layout of a 2-species mixture at Gisburn

4 *Mixture comparisons*
The mixed-species stands consist of a checkerboard planting of groups of 3 x 6 trees of one species, alternating each way with similar groups of the admixed species (Figure 1). Compared with soil differences between monoculture species, those developing between subplots under the various mixture treatments (ie under a given single species within the 18-tree groups) could be expected to be small and possibly undetectable, using traditional chemical methods. Nevertheless, that soil differences might exist was suggested by the presence of differences in tree performance observed in the mixed plots (Lines 1982). Spruce, oak and alder heights have been significantly influenced by the presence of the admixed species (Figure 2). The sequence of effects brought about by the admixed species is pine >alder >oak = spruce (with pine and alder always having a positive

influence, spruce invariably a negative one, and oak a negative or neutral effect). In contrast, height growth of pine has remained identical, whether pure or mixed. One possible explanation of such differential height growth is through changes in soil 'fertility'. As a starting point for differentiating the changes in these soils, the cotton strip assay was again applied. Because of manpower limitations, oak and oak mixtures were not included.

All the strips were buried in June 1980 in each treatment of both pure and mixed trees (except for the different oak combinations) and in the 3 replicate blocks. In the mixed stands, sampling points were located as near as possible to the centre of 8 sample 18-tree 'subplots', distributed on a systematic basis through each plot and in equivalent positions in pure stands. Four strips were inserted, about 20 cm apart, at each sampling point. One strip per sampling position was subsequently removed on each of 4 retrieval occasions, 3, 6, 9 and 12 weeks after insertion. Following standard preparation and testing for tensile strength (Latter & Howson 1977), decay curves were derived (Figure 3), again averaged over the whole strip, and combining data from all 3 blocks.

Differences between tree treatments were significant, and it is evident that the sequence of mixture effects on CTSL for spruce and alder, at least, was the same as that for the height data, ie pine >alder >spruce. That these 2 variables appear to be related is indicated in Figure 4, in which tree heights are plotted against CTSL after 9 weeks, ie the position on the decay curve closest to a 50% CTSL. This degree of CTSL is regarded as providing optimum information (Hill *et al.* 1988). In both spruce and alder, highly significant relationships exist. In the case of pine, in which there was no effect of mixtures on heights, there is a corresponding lack of a mixture effect on CTSL (Figure 3).

How can this close relationship between tree heights and the decomposition of buried cotton cloth be interpreted? One possible hypothesis is that the trees – in mixture – are growing better through some nutritional advantage, as already suggested, and that the increased CTSL in the mixed plots reflects an increase in the availability of the appropriate nutrient(s). The standard approach for checking on the nutritional status of trees – at least in conifers – is to carry out foliar analysis. Where the availability of an element is limiting tree growth, a relationship between foliar concentration of the element in question and tree height growth can sometimes be expected. At Gisburn, foliar sampling of the spruce trees (both pure and mixed) was carried out in the autumn of 1982, taking a lateral shoot from the first whorl on the south side of 10 trees per plot. All 3 replicate blocks were sampled, and the needles were analysed for N, P and K – the 3 nutrients which most commonly limit tree growth on this type of site. No significant relationship between tree heights and K content was obtained, but

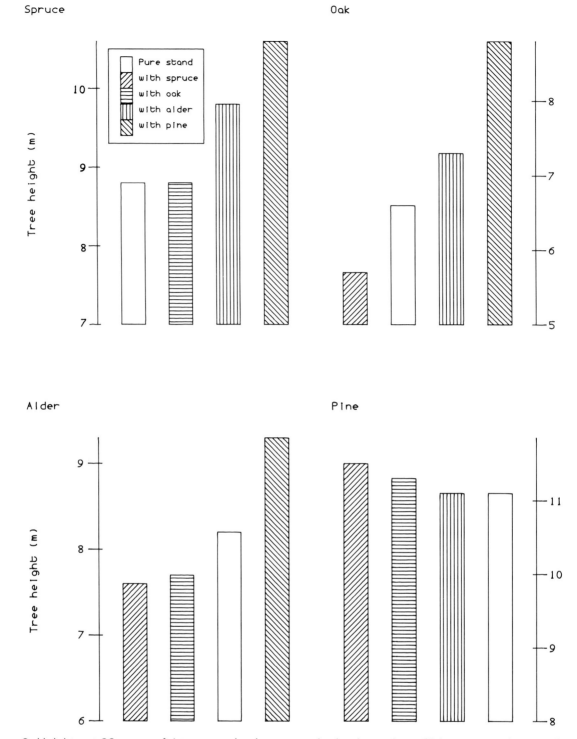

Figure 2. Heights at 26 years of 4 tree species in pure and mixed stands at Gisburn, 1981 (mean of 3 blocks)

the correlations with both N and P concentrations were statistically significant (Figure 5). These graphs make it clear that spruce growing in the presence of pine and alder were not only tallest but had highest foliar N and P, whereas the poorer growth of the monoculture replicates and those mixed with oak was associated with lowest mean foliar concentrations.

The most likely source of these higher levels of N and P in the plots with better growth – which, as has already been noted, were those with the highest CTSL – is the organic matter within and/or beneath the forest floor (Brown & Harrison 1983). The release of N and P, known to be the most tightly held nutrients within organic materials, is likely to vary directly with organic turnover. In fact, a direct relationship between CTSL and foliar N or P can be demonstrated for spruce stands at Gisburn (Figure 6), supporting the view that the improvement in N and P nutrition has been derived largely from soil organic, rather than any other, sources. It should not be inferred from the present results that foliar analysis could replace soil assessment. In the present case, foliar analysis alone would have given no direct clue to the source of any nutrient found

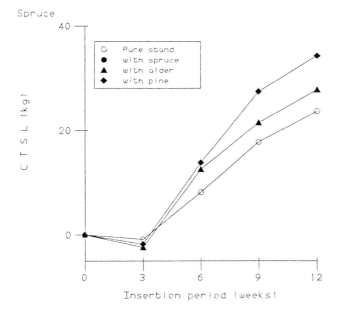

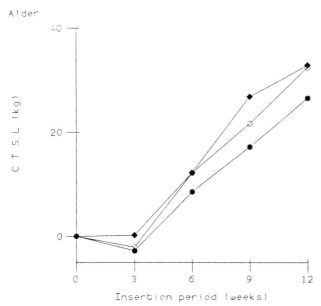

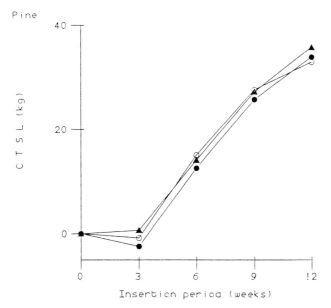

Figure 3. Inverse decay curves for cotton strip tensile strength loss (CTSL) under 3 tree species in pure and mixed stands at Gisburn, 1980 (mean of 3 blocks)

to be limiting growth. Further, in other situations, foliar analysis may be impossible, inappropriate or yield negative results, even where differences in soil properties occur – differences which may well be amenable to assay by the cotton strip method.

5 Conclusion

The evidence from the Gisburn experiment supports the view that the different mixture treatments are influencing N and P availability, presumably from organic matter turnover, and that, in turn, these differences in N and P levels influence tree growth. All these site attributes, which can be loosely related to site 'fertility', are well reflected by the results from the cotton strip assay. The use of this assay, therefore, enabled discrimination to be made, not only between the monocultures despite the lack of detectable soil chemical differences, but also between the different soils developed under a given single tree species, when influenced by the presence of other admixed species.

6 References

Ball, D.F., Dale, J., Sheail, J., Dickson, K.E. & Williams, W.M. 1981. *Ecology of vegetation change in upland landscapes. Part I: General synthesis.* (Bangor occasional paper no. 2.) Bangor: Institute of Terrestrial Ecology.

Brown, A.H.F. 1978. The Gisburn experiment: effects of different tree species on the activity of soil microbes. *Annu. Rep. Inst. terr. Ecol. 1977*, 41.

Brown, A.H.F. & Harrison, A.F. 1983. Effects of tree mixtures on earthworm populations and nitrogen and phosphorus status in Norway spruce (*Picea abies*) stands. In: *New trends in soil biology*, edited by Ph. Lebrun, H.M. André, A. De Medts, C. Grégoire-Wibo & G. Wauthy, 101-108. Ottignies-Louvain-la-Neuve: Dieu-Brichart.

Brown, A.H.F. & Howson, G. 1988. Changes in tensile strength loss of cotton strips with season and soil depth under 4 tree species. In: *Cotton strip assay: an index of decomposition in soils*, edited by A.F. Harrison, P.M. Latter & D.W.H. Walton, 86-89. (ITE symposium no. 24.) Grange-over-Sands: Institute of Terrestrial Ecology.

Harrison, A.F. 1985. Effects of environment and management on phosphorus cycling in terrestrial ecosystems. *J. environ. Manage.*, **20**, 163-179.

Hill, M.O., Latter, P.M. & Bancroft, G. 1988. Standardization of rotting rates by a linearizing transformation. In: *Cotton strip assay: an index of decomposition in soils*, edited by A.F. Harrison, P.M. Latter & D.W.H. Walton, 21-24. (ITE symposium no. 24). Grange-over-Sands: Institute of Terrestrial Ecology.

Holmes, G.D. & Lines, R. 1956. Mixture experiments. *Rep. Forest Res., Lond., 1955*, 34-35.

Howard P.J.A. 1988. A critical evaluation of the cotton strip assay. In: *Cotton strip assay: an index of decomposition in soils*, edited by A.F. Harrison, P.M. Latter & D.W.H. Walton, 34-42. (ITE symposium no. 24.) Grange-over-Sands: Institute of Terrestrial Ecology.

Jacks, G.V. 1963. The biological nature of soil productivity. *Soils Fertil., Harpenden*, **26**, 147-150.

Latter, P.M. & Howson, G. 1977. The use of cotton strips to indicate cellulose decomposition in the field. *Pedobiologia*, **17**, 145-155.

Lines, R. 1982. Species: mixture experiments. *Rep. Forest Res., Edin., 1982*, 13-14.

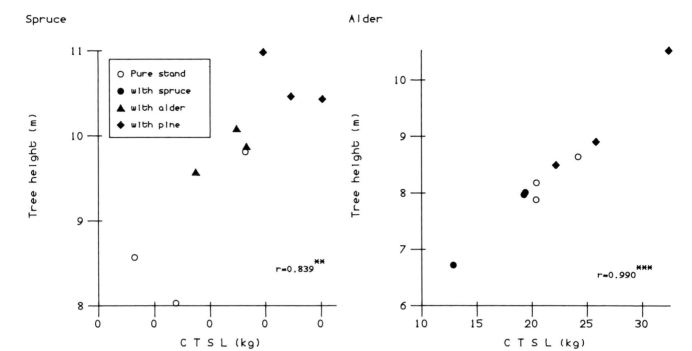

Figure 4. Relationship between tree heights at 26 years (1981) and tensile strength loss of cotton strips after 9 weeks (1980) in pure and mixed stands at Gisburn (means per plot)
(** Significant at P<0.01; *** P<0.001)

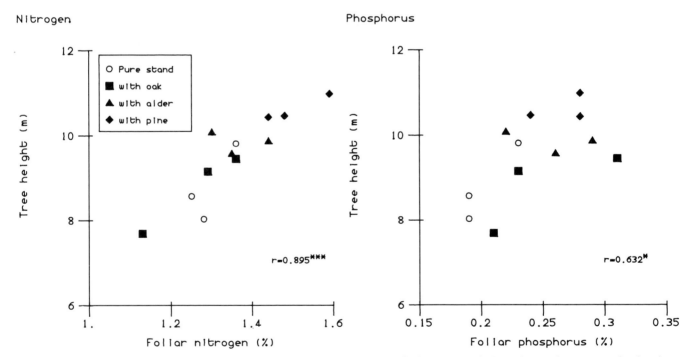

Figure 5. Relationship between tree heights and concentrations of nitrogen and phosphorus in pure and mixed spruce stands at Gisburn, 1981 (means per plot)
(* P<0.05; *** P<0.001)

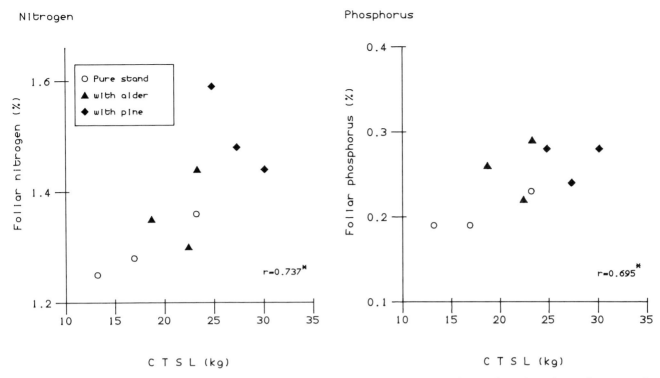

Figure 6. Relationship between foliar concentrations of nitrogen and phosphorus (1981) and tensile strength loss of cotton strips for pure and mixed spruce stands (1980) at Gisburn (means per plot) (P<0.05)

Changes in tensile strength loss of cotton strips with season and soil depth under 4 tree species

A H F BROWN and G HOWSON
Institute of Terrestrial Ecology, Merlewood Research Station, Grange-over-Sands

1 Summary
Seasonal changes in tensile strength loss of cotton strips (CTSL), used as an index of potential organic matter breakdown, were determined in stands of 4 tree species at Gisburn (Bowland Forest), Lancashire.

In general, decomposer potential is greater under alder (*Alnus* spp.) and pine (*Pinus* spp.) than under oak (*Quercus* spp.) and spruce (*Picea* spp.). The seasonal pattern under oak (with a mid-season decline in activity) is very different from the seasonal curves for the other species. CTSL also declines rapidly with depth under oak and spruce, remaining relatively high at all depths under the alder and pine. This higher potential for decomposition at depth under alder and pine is possibly related to the appreciable populations of earthworms (Lumbricidae) under these 2 tree species; conversely, earthworms are rare in oak and spruce stands.

It has also been observed that feeding roots of oak and spruce tend to proliferate near the surface, but are distributed more evenly with depth in alder and pine, these patterns being related to those for CTSL.

2 Background
The background to the study and a description of the site are given by Brown and Harrison (1983) and Brown (1988). Briefly, the Gisburn study area consists of an experimental planting of 4 tree species planted in 0.2 ha plots, both as monoculture stands and in all possible combinations of 2-species mixtures. These 10 treatments are replicated in 3 blocks. The four species are Scots pine (*Pinus sylvestris*), Norway spruce (*Picea abies*), alder (*Alnus glutinosa*) and sessile oak (*Quercus petraea*). The aim of the study was to use the cotton strip assay to test the hypothesis that the 4 monoculture stands were altering the potential for cellulose decomposer activity in the soils in which the biomass of earthworms and the distribution of fine roots were also found to differ.

3 Method
Organic matter decomposer activity is known to vary with both soil depth and season (Swift *et al.* 1979). In using the cotton strip assay under the different tree species, we wished to determine whether there was also any species interaction in cellulose decomposition with either of these sources of variation. Although the standard cotton strip assay gives information on depth differences, it is necessary to repeat the assay at intervals to provide seasonal information. To this end, assays were repeated on 9 oc-casions, at 6-weekly intervals (ie for just over a year) in the monoculture plots of pine, spruce, alder and oak for each of the 3 replicate blocks. Assay periods of 12 weeks were used, thus leading to an overlapping time series of assays. Ten strips per plot were used for each test on each occasion, distributed on a systematic basis throughout each tree plot. The Shirley Soil Burial Test Fabric (1976 batch) was used, the first assay period starting on 5 January 1978, the last finishing on 26 February 1979. The methodology of Latter and Howson (1977) was used throughout, providing tensile strength loss (CTSL) data for 5 depths at 4 cm intervals down the soil profiles. The data were also calculated as cotton rotting rate (CRR), after Hill *et al.* (1988).

Cloth control tensile strength was 47.4 kg for the batch of cloth used on occasions 1–4 and 54.4 kg for the subsequent batch used on occasions 5–9.

To provide comparisons between cotton assay data and other site parameters for the different tree stands, data on earthworm populations and the standing crop of fine roots are also presented. The earthworms were sampled during 1981 using a dung bait method (Brown & Harrison 1983) for 2 of the 3 blocks, but excluding oak stands. Some preliminary data on distribution of live roots ≤ one mm were obtained during 1983 by hand-sorting replicate soil cores from the pine and spruce stands only.

4 Results and discussion
The results from the cotton strip assay indicated that there were clear seasonal patterns of cellulose decomposer activity associated with the different tree species plots (also discussed by French 1988). Because these results were consistent between the separate blocks, the data have been combined from the 3 blocks to give means based on 30 strips per tree species. CTSL or CRR averaged for all 5 depths, ie representing whole strips, provides a summary of the changes in decomposer activity with time in each forest stand (Figure 1).

Statistically significant differences for CTSL ($P<0.001$ on first, second, fifth, sixth occasions, and $P<0.01$ for the other 4 occasions) occurred on all except the last sampling occasion of the series. It is evident that, overall, cellulose decomposer activity was in the sequence alder = pine>oak>spruce. Activity rose to a peak in late summer/early autumn in alder, pine and spruce stands, but with the pattern under oak being very different. Because the successive assay periods

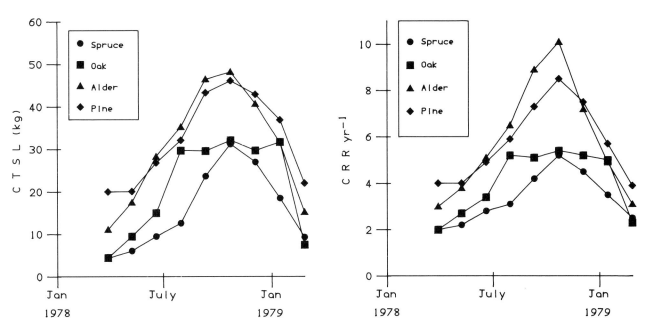

Figure 1. Changes with season in tensile strength loss of cotton strips, CTSL and CRR, following insertion under 4 tree species, at Gisburn, 1978–79. Nine overlapping burial periods, each of 12 weeks; means of all depths (ie whole strip averages) for 3 blocks combined, points being placed on the retrieval date (for a 3D presentation of the oak data, see Walton 1988)

were of identical length, the seasonal curves for CTSL and CRR are essentially similar. Only alder shows a slight discrepancy, when TS values were excessively low in summer (cf Walton 1988).

The seasonal patterns for the separate depths under each tree species (Figure 2) show that the different pattern for oak became more accentuated with depth, being similar to those for other species only at the surface. At 16–20 cm, the oak curve was clearly bimodal, with a secondary minimum at the time at which the other species reached their seasonal maxima.

Inspection of the 4 sets of seasonal curves also reveals that the decline in CTSL with depth, during the season of high activity at least, was slight in alder and pine stands, but very marked under oak and spruce. This contrast between alder and pine, on the one hand, and oak and spruce, on the other, is possibly linked with differences in both earthworm populations and the distribution of the fine (ie the main feeding) roots. Figure 3 shows that alder and pine not only had relatively high numbers of earthworms, but that many of them were *Lumbricus rubellus*, a partially burrowing species. The remainder consisted of surface-dwelling species (Brown & Harrison 1983). In contrast, the spruce stands had very few earthworms and negligible *Lumbricus rubellus*. From observation, populations of earthworms in the oak plots were judged to be at least as low as those of spruce. It is, therefore, possible that there has been more transport both of organic matter and of its decomposer organisms into the soil profiles of alder and pine stands through earthworm activity; and that this is reflected in the relatively high levels of CTSL at all sampled depths.

Conversely, where the possibility of such mixing and transport down the profile has only been slight, as in spruce (and oak) stands, there has been a concomitant rapid fall-off of cellulose decomposer activity down the profile.

Similar reasoning could also explain the contrast in fine root distribution between pine and spruce (Figure 4). Under pine, with little reduction in cellulose decomposer activity (and concomitant nutrient release) with depth, feeding roots likewise tend to occur at all depths, with least at the surface where other environmental conditions, such as susceptibility to drought and temperature fluctuations, are likely to be least suitable for root activity. In contrast, because cellulose decomposer activity in the spruce stands was very much greater at the surface than at depth, the feeding roots were concentrated at the main source of nutrients made available. Because of the resulting surface-rooting of spruce under such conditions, this species is typically more prone to drought than the deeper-rooting pine.

It must be emphasized, however, that these interpretations now require testing in order to distinguish cause from effect.

5 Acknowledgements

Many people assisted in the field work, laboratory testing and data handling involved in this study, including S H Atkinson, A E Elliott, C L Gardener, S Gardener, and R W Hollstein.

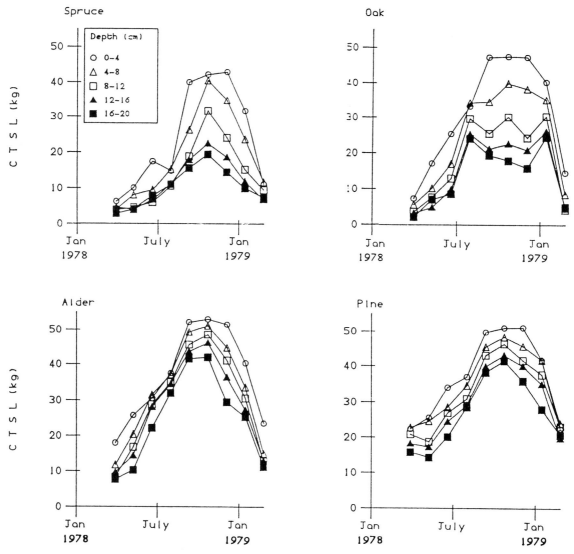

Figure 2. Changes with depth of the seasonal curves for tensile strength loss under 4 tree species

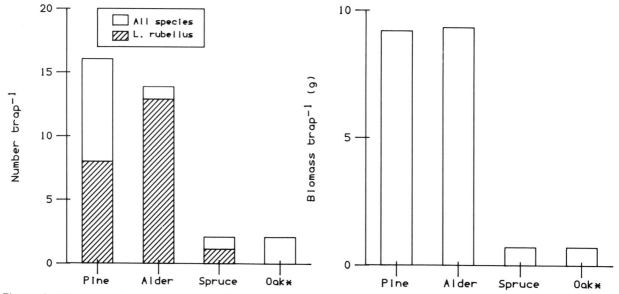

Figure 3. Mean numbers and biomass (preserved fresh weight) of earthworms per trap under 4 tree species. Two trapping periods and 2 blocks, combined, at Gisburn, 1981 (*oak estimated from observation)

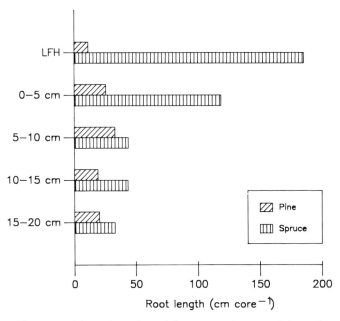

Figure 4. Mean lengths of live roots (≤ 1 mm) in soil cores of 24 mm diameter at 5 depths under Norway spruce (n = 12) and Scots pine (n = 20), at Gisburn, 1983 (source: J M Sykes & S M C Robertson unpublished data)

6   References

Brown, A.H.F. 1988. Discrimination between the effects on soils of 4 tree species in pure and mixed stands using cotton strip assay. In: *Cotton strip assay: an index of decomposition in soils,* edited by A.F. Harrison, P.M. Latter & D.W.H. Walton, 80-85. (ITE symposium no. 24.) Grange-over-Sands: Institute of Terrestrial Ecology.

Brown, A.H.F. & Harrison, A.F. 1983. Effects of tree mixtures on earthworm populations and nitrogen and phosphorus status in Norway spruce (*Picea abies*) stands. In: *New trends in soil biology,* edited by Ph. Lebrun, H.M. André, A. De Medts, C. Grégoire-Wibo & G. Wauthy, 101-108. Ottignies-Louvain-la-Neuve: Dieu-Brichart.

French, D.D. 1988. Seasonal patterns in cotton strip decomposition in soils. In: *Cotton strip assay: an index of decomposition in soils,* edited by A.F. Harrison, P.M. Latter & D.W.H. Walton, 46-49. (ITE symposium no. 24.) Grange-over-Sands: Institute of Terrestrial Ecology.

Hill, M.O., Latter, P.M. & Bancroft, G. 1988. Standardization of rotting rates by a linearizing transformation. In: *Cotton strip assay: an index of decomposition in soils,* edited by A.F. Harrison, P.M. Latter & D.W.H. Walton, 21-24. (ITE symposium no. 24.) Grange-over-Sands: Institute of Terrestrial Ecology.

Latter, P.M. & Howson, G. 1977. The use of cotton strips to indicate cellulose decomposition in the field. *Pedobiologia,* **17,** 145-155.

Swift, M.J., Heal, O.W. & Anderson, J.M. 1979. *Decomposition in terrestrial ecosystems.* Oxford: Blackwell Scientific.

Walton, D.W.H. 1988. The presentation of cotton strip assay results. In: *Cotton strip assay: an index of decomposition in soils,* edited by A.F. Harrison, P.M. Latter & D.W.H. Walton, 28-31. (ITE symposium no. 24.) Grange-over-Sands: Institute of Terrestrial Ecology.

Effects of tree species on forest soils in northern France, detected by cotton strip assay

C NYS[1] and G HOWSON[2]

[1]Centre de Recherche Forestiere (CNRF), Champenoux, France
[2]Institute of Terrestrial Ecology, Merlewood Research Station, Grange-over-Sands

1 Summary

The aim of the study was to determine whether or not the cotton strip assay is sufficiently sensitive to act as an indicator of changes in soil properties induced by forest trees.

Three forest sites were used for these studies: one in the Ardennes, with sessile oak (Quercus petraea) or Norway spruce (Picea abies), on an acid brown earth overlying shale; the second in Normandy with oak, beech (Fagus sylvatica) or spruce, on a leached brown earth formed from loess overlying clay with flints; and the third in the Vosges, with Scots pine (Pinus sylvestris), on a podzolic soil over grit. After a pilot study, cotton strips were inserted in soil profiles in the winter (December–February) for 8 weeks and in summer (July) for 3 weeks during 1983–84.

The results show that the cotton strip assay is sensitive and is able to reflect changes in some organic and physical properties. Between-forest site, between-species, between-soil depth and seasonal effects on cotton strip decomposition were detected. The between-species effects were detected in the surface soil layer only in the summer period, and were of the same order for the 2 sites where comparisons were possible. It was concluded that the seasonal factor should be taken into account in studying tree species effects on soils, and that the study should be extended to other sites.

2 Introduction

Trees and associated forest management practices are known to cause changes in soil properties and fertility: some changes induced are of low intensity and are often detectable only after a long period (Duchaufour & Bonneau 1961; Bollen 1974; Page 1974; Bonneau et al. 1976; Nys 1981). At CNRF, we are currently studying changes in the quality of organic matter, microbial activity and chemical and physical properties of soils brought about by different tree species (Nys 1981; Bonnaud et al. 1985; Nys & Ranger 1985; Nys et al. 1987).

The rate at which cellulose decays in soil differs in various forest sites and at the same site under different tree species (Lahde 1974; Berg et al. 1975). Change in tensile strength of cotton strips after insertion in soil has been proposed as an index of the rate of cellulose decomposition (Latter & Howson 1977), and the rate of decomposition of these strips has been found to vary in soil under different tree species. The rates have also been shown to correlate well with the improvement in tree growth in species mixtures (Brown & Harrison 1983; Brown & Howson 1988).

The aim of the present study was to test whether or not the cotton strip assay could be a sensitive indicator of changes in potential decomposition, within the framework of our studies on the effect of forest trees on soil fertility in northern France.

3 Site description

The 3 sites are situated in Normandy, the Ardennes and the Vosges, in France. The first 2 sites have sessile oak as climax species and a similar range of other species. The third site, in the Vosges, is under Scots pine.

3.1 Normandy site

The Normandy site is situated at Eu forest, 150 km west of Paris, at an elevation of 200 m. The soil (Nys et al. 1987) is an eluviated brown earth (sol brun lessive); some properties are given in Table 1. The

Table 1. Characteristics of organic matter in surface litter and A_1 horizon (organo mineral horizon)

| | Normandy | | | Ardennes | | Vosges |
	Oak	Beech	Spruce	Oak	Spruce	Pine
Litter (t ha⁻¹)	14.0	26.8	54.2	17.3	37.3	
A_1 horizon						
Carbon (t ha⁻¹)	23.1	18.3	33.0	56.6	81.5	33.1
Nitrogen (t ha⁻¹)	1.8	1.3	1.5	3.9	4.8	1.1
C/N	12.8	14.1	22.0	10.5	16.4	32.7
Organic matter (% C total)						
Fresh	4.7	6.9	12.7			
Bound	41.0	24.5	8.2			
Residual humin	1.9	6.0	6.7			
pH	4.4	4.5	4.0	4.2	3.7	3.4
Humus	Acid mull	Acid mull	Moder	Acid mull	Moder	Dysmoder

soils are formed from a loess deposit, one to several metres thick, overlying a layer of clay-with-flints on cretaceous chalk. The climate is oceanic, with a mean annual rainfall of 777 mm evenly distributed throughout the year and a mean annual temperature of 9.5°C.

Oak: A sessile oak stand about 100 years old, with a mean tree diameter at breast height (dbh) of 35 cm, this site was used as the initial forest control plot. The ground flora consists of grasses, cow-wheat (*Melampyrum* spp.), bracken (*Pteridium aquilinum*) and bramble (*Rubus* spp.).

Beech: A stand of beech (*Fagus sylvatica*), 150–200 years old, now about 35 m high, with a dbh of 55 cm, and virtually no ground flora.

Spruce: Norway spruce, about 80 years old, with a density of 150 trees ha^{-1}, 35–40 m high, and a dbh of 60 cm. The ground flora is mainly bracken.

3.2 Ardennes site

The Ardennes site is situated at Montherme, 200 km east of Paris, at an elevation of 400 m. The soil is a brown earth (sol brun acide) (Table 1), formed from a silt deposit overlying a revinien shale (Nys 1981). The mountain climate has a mean annual rainfall of 1300 mm, distributed throughout the year, and a mean annual temperature of 8.5°C.

Oak: A coppice with sessile oak standards, with 70 trees ha^{-1}, 150–200 years old, and 17 m high. The stand consists of 30-year-old oak (35%), birch (*Betula verrucosa*) (40%) and rowan (*Sorbus aucuparia*) (20%). The ground flora consists of grasses, bracken and bilberry (*Vaccinium myrtillus*).

Beech: Isolated trees of beech, 10 ha^{-1}, 150–200 years old, 17–20 m high covering 100 m^2 each. There is no ground flora.

Spruce: The Norway spruce were planted 50 years ago after clearfelling of part of the oak stand. The trees are 20–22 m high, with a dbh of 25 cm and 900 trees ha^{-1}. There is no ground flora.

3.3 Vosges site

The Vosges site is situated at Grandviller, 350 km east of Paris, at an elevation of 400 m, and the soil is a podzolic soil on grit. The climate is semi-continental, with a mean annual rainfall of 900 mm. Scots pine trees were planted 30 years ago and, after thinning, there are now 700 trees ha^{-1}, 15 m high.

4 Method

After a pilot study in the summer of 1983, 10 cotton strips were inserted in areas between trees adjacent to the litter traps, which were already on the plots.

Methods for assessing the organic matter properties of the forest soils (Table 1) are given in Bonneau and

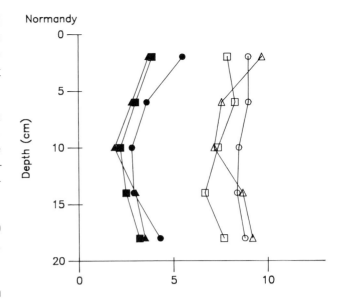

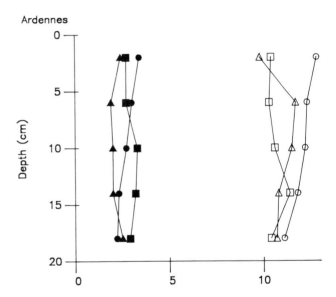

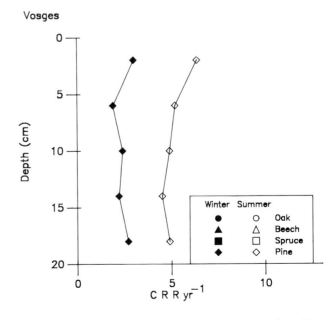

Figure 1. Rotting rates (CRR) of cotton strips. The variation between tree species and season in 3 sites in France

Table 2. Analysis of variance of cotton strip rotting rates (CRR yr^{-1}); F values for site and species effects

	Depth (cm)				
	0–4	4–8	8–12	12–16	16–20
Winter 1983					
n = 89					
Site	5.5 **	1.5 NS	0.25 NS	0.4 NS	4.5 *
Species	1.5 NS	0.6 NS	1.2 NS	0.6 NS	1.5 NS
Site x species	0.9 NS	0.5 NS	2.1 NS	3.4 NS	2.6 NS
Summer 1984					
n = 107					
Site	5.8 **	9.1 **	15.7 **	12.7 **	4.8 **
Species	2.4 *	0.9 NS	0.9 NS	0.9 NS	1.0 NS
Site x species	5.5 *	1.4 NS	0.3 NS	2.0 NS	0.3 NS

NS, not significant; * P<0.05; ** P<0.01

Table 3. Analysis of variance of cotton strip rotting rates (CRR yr^{-1}); mean ± SD; F values for site effect for means of all species

	Depth (cm)				
	0–4	4–8	8–12	12–16	16–20
Winter 1983					
Normandy	4.60 ± 1.68	3.10 ± 1.46	2.56 ± 1.29	2.67 ± 1.21	3.29 ± 1.39
Ardennes	2.70 ± 1.24	2.48 ± 1.24	2.45 ± 1.21	2.37 ± 0.88	2.56 ± 0.91
Vosges	2.92 ± 0.69	1.79 ± 0.73	2.30 ± 1.02	2.15 ± 0.77	2.30 ± 0.69
F ratio	14.65**	4.7*	0.2 NS	1.3 NS	4.3*
Summer 1984					
Normandy	9.02 ± 2.7	7.99 ± 2.9	7.64 ± 2.7	7.65 ± 2.7	8.11 ± 3.2
Ardennes	11.12 ± 2.7	11.50 ± 2.6	11.15 ± 2.2	11.07 ± 2.5	11.11 ± 2.1
Vosges	6.35 ± 1.8	5.20 ± 1.7	4.92 ± 2.1	4.50 ± 1.7	4.91 ± 2.5
F ratio	13.6**	28.1**	33.6**	32.6**	22.9**

NS, not significant; * P<0.05; ** P<0.01

Souchier (1982). The winter series were inserted for 8 weeks and the summer series for 3 weeks, from December 1983–February 1984 and in July 1984, respectively. Cotton strips were inserted and processed according to the method of Latter and Howson (1977), and the data expressed as cotton rotting rate (CRR) according to Hill *et al.* (1988).

5 Results

The rates of decomposition of cotton strips (CRR) decreased in the order oak>beech>spruce at both sites (Figure 1), but the species effect is only significant in the humus layer of the soil (Table 2). Figure 1 also shows that season, under the French climate, had the greatest effect on cellulose decomposition, significant at 1% (F ratio = 275, df 1, 194). The site effect shows different decomposition rates (P<0.01) in the summer (Table 3), despite the dry summer of 1984.

A good relationship is apparent between the order of the CRR and the classification of humus type and some soil properties over the 6 plots examined, the CRR values increasing with increasing pH, decreasing carbon/nitrogen (C/N) and decreasing quantity of litter across the sites. Thus, beech or oak sites with lower quantity of litter and C/N ratio and higher pH showed higher cellulose decomposition rates than the spruce plots. Figure 2 shows, for the Normandy site, that the physical properties of soils are also highly related with the CRR. The better the soil porosity, clay stability and soil aggregation, the faster is the decomposition of the cotton strips.

6 Conclusion

The experiment indicates that the assay was sufficiently sensitive to differentiate between the effects on soils of different tree species, despite the dry summer of 1984. The relationships between cotton strip decomposition rates and humus form and soil physical and chemical properties show that this method could be used as a test of soil biological activity in the same way that the mineral bag method is used as an indicator of decline in soil fertility (Bonnaud *et al.* 1985).

It is concluded that the experiment should be repeated at more sites with a similar range of tree species or silvicultural treatments, because of the very significant interaction between sites and species. Cotton strips should also be inserted for different periods of the year to examine the seasonal variation patterns (cf Brown & Howson 1988) for different sites.

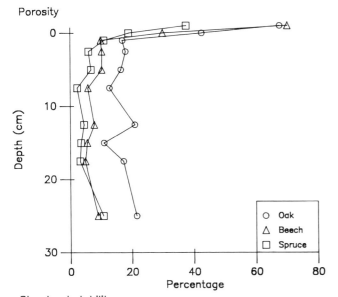

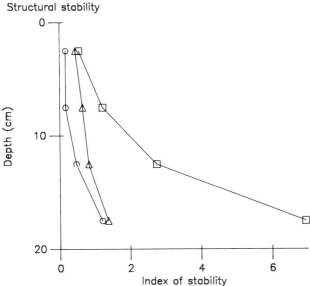

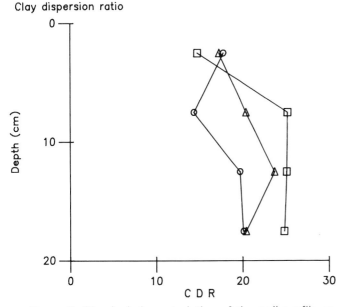

Figure 2. Physical characteristics of the soil profile at the Normandy site

   i. Porosity is the area of pores/total analysed area, expressed as a percentage

  ii. Structural stability (method: Henin et al. 1969)

 iii. Clay dispersion ratio (method: MAFF 1982)

7   References

Berg, B., Karenlampi, L. & Veum, A.K. 1975. Comparisons of decomposition rates measured by means of cellulose. In: *Fennoscandian tundra ecosystems, Part 1. Plants and microorganisms*, edited by F.E. Wielgolaski, 261-267. Berlin: Springer.

Bollen, W.B. 1974. Soil microbes. In: *Environmental effects of forest residues management in the Pacific Northwest*, edited by O.P. Cramer, B1-41. (USDA Forest Service Gen. Tech. Rept PNW-24.) Portland, Oregon: US Department of Agriculture.

Bonneau, M. & Souchier, B. 1982. *Constituents and properties of soils.* London: Academic Press.

Bonneau, M., Brethes, A., Nys, C. & Souchier, B. 1976. Influence d'une plantation d'epiceas sur un sol du Massif central. *Lejeunia, N.S.,* no. 82, 1-14.

Bonnaud, P., Hatton, A., Nys, C., Ranger, J. & Robert, M. 1985. Soil physico-chemical changes studied by the *in situ* weathering of an introduced primary mineral. In: *1st IUFRO Workshop on qualitative and quantitative assessment of forest sites with special reference to soil.* Birmensdorf: W. Bosshard, Swiss Federal Institute of Forest Research.

Brown, A.H.F. & Harrison, A.F. 1983. Effects of tree mixtures on earthworm populations and nitrogen and phosphorus status in Norway spruce (*Picea abies*) stands. In: *New trends in soil biology,* edited by Ph. Lebrun, H.M. André, A. De Medts, C. Grégoire-Wibo & G. Wauthy, 101-108. Ottignies-Louvain-la-Neuve: Dieu-Brichart.

Brown, A.H.F. & Howson, G. 1988. Changes in tensile strength loss of cotton strips with season and soil depth under 4 tree species. In: *Cotton strip assay: an index of decomposition in soils,* edited by A.F. Harrison, P.M. Latter & D.W. Walton, 86-89. (ITE symposium no. 24.) Grange-over-Sands: Institute of Terrestrial Ecology.

Duchaufour, Ph. & Bonneau, M. 1961. Evolution d'un sol de foret feuillue 'Terra fusca' provoqué par une plantation de Douglas *Pseudotsuga douglasii* d'une trentaine d'annes. *Revue for. fr.,* **12,** 793-799.

Henin, S. Gras, R. & Monnier, 1969. *Le profil cultural.* Paris: Masson et Cie.

Hill, M.O., Latter, P.M. & Bancroft, G. 1988. Standardization of rotting rates by a linearizing transformation. In: *Cotton strip assay: an index of decomposition in soils,* edited by A.F. Harrison, P.M. Latter & D.W. Walton, 21-24. (ITE symposium no. 24.) Grange-over-Sands: Institute of Terrestrial Ecology.

Lahde, E. 1974. Rate of decomposition of cellulose in forest soils in various parts of the Nordic countries. *Rep. Kevo Subarct. Res. Stn,* **11,** 72-78.

Latter, P.M. & Howson, G. 1977. The use of cotton strips to indicate cellulose decomposition in the field. *Pedobiologia,* **17,** 145-155.

Ministry of Agriculture, Fisheries and Food. 1982. *Techniques for measuring soil physical properties.* (Reference book 441.) London: HMSO.

Nys, C. 1981. Modifications des caracteristiques physico-chimiques d'un sol brun acide des Ardennes primaires par la monoculture d'Epicea commun. *Ann. Sci. for.,* **38,** 237-258.

Nys, C. & Ranger, J. 1985. Influence de l'espece forestiere sur le fonctionnement de l'ecosysteme forestier. Le cas de la substitution d'une essence resineuse a une essence feuillue. *Science Sol,* **4,** 203-216.

Nys, C., Bullock, P. & Nys, A. 1987. Micromorphological and physical properties of soil under three different species of trees. In: *Proc. 7th int. meeting on soil micromorphology,* edited by N. Fedoroff & L.M. Bresson. Paris: Institut National Agronomique – Association Francais pour l'Etude du Sol.

Page, G. 1974. *Effects of forest cover on the properties of some Newfoundland soils.* (Canadian Forestry Service publication no. 1332.) Ottawa: Department of the Environment.

Use of the cotton strip assay to detect potential differences in soil organic matter decomposition in forests subjected to thinning

G HOWSON

Institute of Terrestrial Ecology, Merlewood Research Station, Grange-over-Sands

1 Summary

The cotton strip assay was used to detect potential changes in rates of soil organic matter decomposition associated with different thinning treatments of Sitka spruce (*Picea sitchensis*) stands at Elwy Forest, north Wales, UK. Cotton strips were inserted in the soil over a period of one year, in 4 3-month periods. In general, the losses in the tensile strength (CTSL), used as an indicator of decomposition of the cotton strip, were greater at the surface than lower in the soil profile. With respect to the forest thinning treatments, the CTSL values were in the order: intermediate thinned >high intensity thinned>unthinned. Higher CTSL values were also detected for the so-called 'tree' areas than for the 'stump' areas within the thinned stands.

The cotton strip data are considered in relation to (i) the amounts of litter accumulated and the annual litterfall under the stands, and (ii) the possible significance of differences in soil organic matter decomposition rates induced by the thinning practice.

2 Introduction

The importance of leaf and branch litter decomposition and consequent nutrient release for the growth of forests ecosystems has long been recognized (Duvigneaud & Denaeyer de Smet 1970; Witkamp & Ausmus 1976; Ausmus *et al.* 1976; Miller *et al.* 1979; Harrison 1978, 1985). Accumulations of litter frequently occurring on the forest floor (Turner & Long 1975; Adams & Dickson 1973; Carey *et al.* 1982) are usually due to imbalances between the rates of production and decomposition of this litter, resulting in a slowing of nutrient cycling (Malmer 1969). Slowing of nutrient cycling in forests, particularly those on poor soils, is likely to result in a reduction of tree productivity. Under those circumstances, therefore, any forest management practice which indirectly stimulates organic matter decomposition, and thus nutrient cycling, would be beneficial to timber production.

Thinning of the forest is known to stimulate the production and timber quality of plantation forests (Assmann 1970; Hibberd 1986). It also appears to result in a reduction in the amount of litter accumulated on the forest floor, though high thinning treatments may result in high litter accumulation due to the brash left behind. A reduction in the amount of litter on the forest floor could be for 2 reasons, either a decrease in the rate of litterfall or an increase in the rate of the litter decomposition. As a preliminary test of the idea that thinning may indirectly stimulate the

decomposition of the litter, the cotton strip assay was applied to a Forestry Commission thinning experiment, for which some data on litter accumulation and litterfall were available.

3 Method

The site selected for study was a Forestry Commission Sitka spruce plantation (Elwy), near St Asaph, north Wales (National Grid reference SJ 078767), planted in 1952 at an altitude of 260 m on an unploughed upland brown earth (brown podzolic soil *sensu* Avery 1980), with some evidence of gleying of the profile at 5–10 cm depth. The thinning treatments were arranged in a randomized block design with 3 replicates, each plot being about 0.5 ha. Line thinning was the method applied, which followed along the planting rows, and was carried out at 3 intensities: (i) unthinned; (ii) intermediate with one line in 3 removed; and (iii) high with 2 lines in 4 removed. The percentage of wood removed was approximately 0, 33 and 50 respectively. The thinning took place in 1971–72, although a few individual trees have been removed subsequently in 1978.

In July 1979, 20 cotton strips were randomly inserted in a vertical manner (Latter & Howson 1977) into the soil, in only 2 of the 3 experimental blocks. Ten were placed within the lines of the remaining standing trees 'tree' and 10 in an adjacent position within the lines of thinned stumps 'stump', as illustrated in Figure 1. Ten strips only were inserted in the soil of unthinned stands. The strips were carefully removed after 3 months, and on the day of removal new replicate strips were inserted in similar adjacent positions. The pattern of insertion and removal was continued for one year, giving 4 3-month periods. Each time, both cloth and field controls were used.

The tensile strength of the strips, prepared as described in Latter and Howson (1977), was measured using a Monsanto Type W tensometer with pneumatic jaws and an automatic chart recorder. The Shirley Soil Burial Test Fabric (1976 batch) was used, and it showed considerable variation in cloth control tensile strength values depending on the roll used. Thus, the field controls had mean tensile strength values of 55.2 kg for insertion periods 1, 2 and 3, and 44.5 kg for period 4. The results presented were adjusted to allow for this variation.

Litterfall was monitored monthly for a whole year from mid-1979, using 8 randomly placed 25 cm

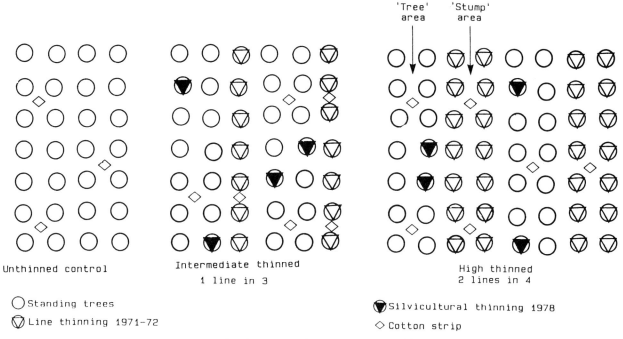

Figure 1. Within thinning treatment plot layout and cotton strip positions

diameter funnel-type litter traps, emptied at monthly intervals. Litter accumulation on the forest floor was recorded in July 1978, by removing 25 10 cm diameter cores per plot from each of the 3 blocks. For litterfall and for litter accumulation, the material was dried at 105°C and weighed. Weights were multiplied up accordingly to give quantities in t ha^{-1}.

4   Results

4.1   Cotton strip assay

Generally, the patterns of tensile strength loss showed that potential decomposition was highest at the soil surface and declined with depth (Figure 2 i).

The unthinned plots appeared to have the lowest CTSL at all the soil depths.

To investigate the effects of the thinning treatments further, a one-way analysis of variance was carried out on the CTSL data for each of the 4 3-month periods separately, using the mean values of the 5 depths from each strip, and pooling the values from the 2 blocks and the 'stump' and 'tree' areas. Bartlett's test (Snedecor & Cochrane 1967) indicated that the within-treatment variances were homogeneous and, therefore, the data needed no transformation. Partitioning the treatment sums of squares showed that unthinned

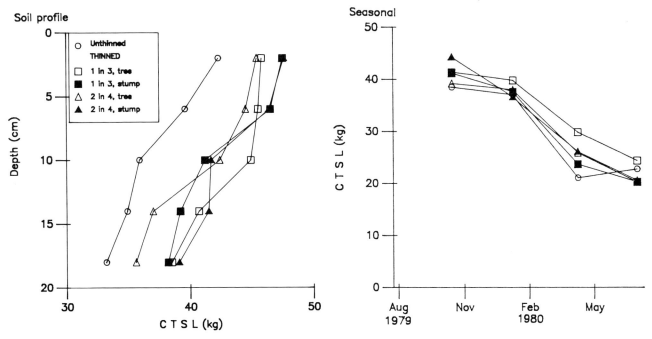

Figure 2. Tensile strength loss of cotton strips (CTSL) in thinning treatment plots

i. change down profile for one block, autumn 1979

ii. change throughout season, means of 2 blocks for the 5 treatments. Points are placed at retrieval dates

plots had significantly lower CTSL values than the combined thinned plots in July–October 1979 (P <0.05) and in January–April 1980 (P<0.01), but that there was no significant difference between the intermediate and high thinned treatments (Table 1). A multiple comparison of means using Tukey's Honestly Significant Difference (HSD) test then revealed that the unthinned plots had a significantly lower CTSL than both the high thinned (P<0.05) and intermediate thinned (P<0.01) treatments, in January–April 1980. In addition, the strips inserted in the 'tree' and 'stump' areas gave significantly different (P<0.05) CTSL values for each depth (Table 1). Clearly, therefore, there appears to be an interactive effect on potential decomposition between the thinning treatments themselves and the zonation within the thinned plots. There was a further subordinate interaction of the seasonal factor with a higher rate of decomposition of strips in the 'tree' position and a lower rate for those in the 'stump' position for the period January–July, the reverse situation occurring in the period July–October (Table 1).

Table 1. Cellulose decomposition in the thinned and unthinned plots and in the 'tree' and 'stump' areas of the thinned plots. The mean is for 5 depths for each strip, both blocks combined
  i. The difference between thinned and unthinned plots is shown by the F ratio obtained from a one-way analysis of variance
  ii. The difference between 'tree' and 'stump' areas is shown by a paired T test

Period	Mean tensile strength loss (kg)			
	Jul–Oct 1979	Oct–Jan	Jan–Apr 1980	Apr–Jul
i.				
Unthinned	14.1	15.5	31.4	21.7
Thinned	11.0	14.5	26.1	23.2
F ratio, df 1, 97	4.6*	0.7	9.8**	0.9
ii.				
Intermediate thinned				
Tree	41.4	39.8	29.9	24.4
Stump	41.2	37.6	23.7	20.3
T test	−0.1	−1.5	−2.6*	−2.2*
High thinned				
Tree	39.2	38.0	25.5	20.2
Stump	44.3	36.6	26.1	20.6
T test, df 19	2.8*	−1.1	0.08	0.2

*P<0.05; **P<0.01

Nevertheless, the intermediate thinned plots appear to provide the best conditions for decomposition, because this treatment showed a higher CTSL over the year as a whole than either the high thinned or unthinned treatments.

4.2 Litterfall
The annual litterfall (Table 2 & Figure 3) was appreciably greater in the unthinned plots than in either of the thinned treatments. The 2 thinned plots had similar litterfall values, despite the different levels of thinning

Table 2. Litterfall and forest floor accumulation for thinned and unthinned areas. Litter weight ha^{-1} with SE for treatment differences

	Thinning treatment			
	Un-thinned	Inter-mediate	High	SE
July 1979–June 1980[1]				
Litterfall (t ha^{-1} yr^{-1})	4.86	2.81	3.01	0.24*
July 1978[2]				
Litter accumulated on the forest floor (t ha^{-1})				
L horizon	14.39	11.20	13.19	1.32 NS
F + H	122.06	92.07	117.34	112.90 NS

* P<0.05
[1] Data supplied by D Evans, ITE, Bangor
[2] Data supplied by Forestry Commission

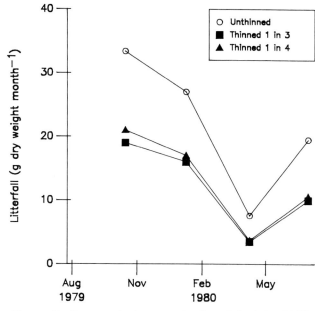

Figure 3. Seasonal changes in litterfall, July 1979–June 1980. Points are mean litterfall per month over 4 3-monthly periods for 2 blocks combined

intensity. A one-way analysis of variance was carried out on the annual litterfall data, combining the 2 blocks but with no data transformation, as Bartlett's test again showed that variances were homogeneous. Tukey's Honestly Significant Difference test then showed that the annual litterfall under the unthinned plot was significantly higher (P<0.05) than under either of the 2 thinned plots.

4.3 Litter accumulation
The quantity of litter (L) and fermentation and humus (F + H) material accumulated on the forest floor (Table 2 & Figure 4) tended to be highest in the unthinned plot, with slightly less material on the high thinned plot and least on the intermediate thinned plot. However, as these data were very limited in replication, the differences were not found to be statistically significant.

4.4 Decay rate
The data for annual litterfall (AL) and for the accumulated litter in the L layer (x) were used to calculate k

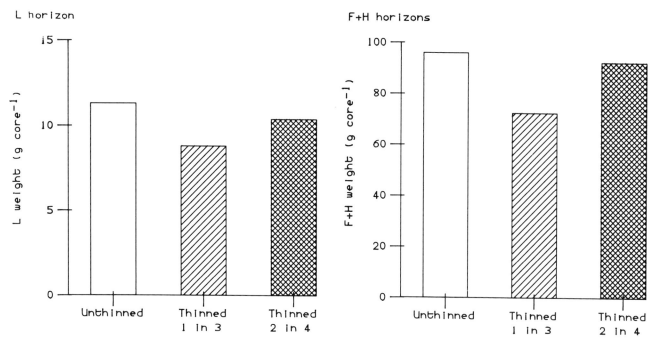

Figure 4. Forest floor accumulation, July 1978, calculated from 25 samples per plot

values and 95% turnover times (ie 3/k years), according to Olsen (1963) and Whittaker (1975), where k = AL/x, and a steady state is assumed (Table 3). The k values were in the order: unthinned>intermediate >high thinned, and the same relative order between treatments was obtained if the F+H and L data were added together.

Table 3. k values and 95% turnover rates for the L layer

Thinning treatment	k	95% time
Unthinned	0.34	9
Intermediate	0.25	12
High	0.23	13

5   *Discussion*

The results show that the cotton strip assay is able to detect significant differences in the potential rate of organic matter decomposition induced in the forest soils by the forest thinning treatments. Furthermore, the assay results indicate that quite complex interactions occur between organic matter decomposition processes and the environmental conditions related to the thinning pattern, soil depth and seasonal factors. These interactions appear to be biologically significant, as the same type of effects have been found in other forest stands (Brown & Howson 1988; Nys & Howson 1988; Brown 1988). Also, the rates of decomposition of the cotton strips directly relate to the soil organic matter decay rates calculated as k (Table 3). Using filter paper, Fox and Van Cleve (1983) have similarly demonstrated that differences in annual rates of cellulose decomposition among stands in an Alaskan taiga forest were correlated with the k values. The changes in the environmental conditions which follow the removal of trees during thinning are complex. Unthinned plots tend to be cooler and drier than thinned plots. Any thinning treatment results in less interception of the rain so that more rainfall reaches the ground, and line thinning may allow more rainfall to reach the ground than selective thinning. Further, the increased throughfall after line thinning tends to exceed any water loss by evaporation (Hamilton 1980), particularly if brash is left on the soil surface, and this, together with lower rates of water uptake by roots and transpiration by fewer trees, results in moister soil conditions. We did observe that cotton strips retrieved from 'stump' areas were wetter than those from the 'tree' areas throughout the year. Everett and Sharrow (1985) also found that soil water content was relatively greater in areas around cut stems of pinyon pine (*Pinus cembroides*) trees than on unthinned plots, and the difference could be detected for up to 4 years. In an experiment on 70-year-old white spruce (*Picea glauca*), near Fairbanks, Alaska, Piene and Van Cleve (1978), using litter and cellulose bags, found that there were differences in decomposition and weight loss influenced by the changes in a number of factors following thinning of plots. Similarly, light intensity also increases following thinning, so temperature changes, too, are likely to be greater in thinned stands, and the 'stump' areas in particular, than in unthinned stands. It is, therefore, realistic to expect that the microbially mediated organic matter decomposition processes would be increased under the improved moisture and temperature conditions provided in the thinned stands, as has been indicated by the cotton strip assay; the synthesis of the collated data (Ineson *et al.* 1988) has shown that the factors with the greatest influence on decomposition rates in cotton strips are temperature and moisture. Though the study has been carried out in only one forest experiment, it appears that the tendency for greater litter accumulation under unthinned stands is not just a result of higher litterfall, but also due to generally slower decomposition rates. The cotton strip assay not only detected this general trend for slower decomposition in the unthinned stands, but has also produced evidence for quite subtle interactions between organic matter decomposition processes in forest soils and environmental factors. In view of the direct links between organic matter turnover, nutrient cycling and tree productivity on the poor afforested upland soils (Miller et al. 1979), these interactions and the potential effects of management practices on them warrant further study.

6 References

Adams, S.N. & Dickson, D.A. 1973. Some short-term effects of lime and fertilizers on a Sitka spruce plantation. *Forestry,* **46,** 31-37.

Assmann, E. 1970. *The principles of forest yield study.* Oxford: Pergamon.

Ausmus, B.S., Edwards, N.T. & Witkamp, M. 1976. Microbial immobilization of carbon, nitrogen, phosphorus and potassium; implications for forest ecosystem processes. In: *The role of terrestrial and aquatic organisms in decomposition processes,* edited by J.M. Anderson & A. Macfadyen, 397-416. Oxford: Blackwell Scientific.

Avery, B.W. 1980. *Soil classification for England and Wales* (Higher categories). (Soil Survey technical monograph no. 14.) Harpenden: Soil Survey of England and Wales.

Brown, A.H.F. 1988. Discrimination between the effects on soils of 4 tree species in pure and mixed stands using cotton strip assay. In: *Cotton strip assay: an index of decomposition in soils,* edited by A.F. Harrison, P.M. Latter & D.W.H. Walton, 80-85. (ITE symposium no. 24.) Grange-over-Sands: Institute of Terrestrial Ecology.

Brown, A.H.F. & Howson, G. 1988. Changes in tensile strength loss of cotton strips with season and soil depth under 4 tree species. In: *Cotton strip assay: an index of decomposition in soils,* edited by A.F. Harrison, P.M. Latter & D.W.H. Walton, 86-89. (ITE symposium no. 24.) Grange-over-Sands: Institute of Terrestrial Ecology.

Carey, M.L. & Farrell, E.P. 1978. Production, accumulation and nutrient content of Sitka spruce litterfall. *Irish For.,* **35,** 35-44.

Carey, M.L., Hunter, I.R. & Andrew, I. 1982. *Pinus radiata* forest floors: factors affecting organic matter and nutrient dynamics. *N.Z. Jl For. Sci.,* **12,** 36-48.

Duvigneaud, P. & Denaeyer de Smet, S. 1970. Biological cycling of minerals in temperate deciduous forests. In: *Analysis of temperate forest ecosystems,* edited by D.E. Reichle, 199-225. London: Chapman & Hall.

Everett, R.L. & Sharrow, S.H. 1985. *Soil water and temperature in harvested and non harvested pinyon-juniper stands.* (Research paper INT-342.) Ogden, Utah: United States Department of Agriculture, Forest Service, Intermountain Research Station.

Fox, J.D. & Van Cleve, K. 1983. Relationship between cellulose decomposition, Jenny's k, forest floor nitrogen and soil temperature in Alaskan taiga forests. *Can. J. For. Res.,* **13,** 789-794.

Hamilton, G.J. 1980. *Line thinning.* (Forestry Commission leaflet no. 77.) London: HMSO.

Harrison, A.F. 1978. Phosphorus cycles of forest and upland grassland ecosystems and some effects of land management practices. In: *Phosphorus in the environment: its chemistry and biochemistry,* edited by R. Porter & D.W. Fitzsimons, 175-199. (Ciba Foundation symposium no. 57.) Amsterdam: Elsevier.

Harrison, A.F. 1985. Effects of environment and management on phosphorus cycling in terrestrial ecosystems. *J. environ. Manage.,* **20,** 163-179.

Hibberd, B.G. 1986. *Forest practices.* (Forestry Commission bulletin no. 14. 10th ed.) London: HMSO.

Ineson, P., Bacon, P.J. & Lindley, D.K. 1988. Decomposition of cotton strips in soil: analysis of the world data set. In: *Cotton strip assay: an index of decomposition in soils,* edited by A.F. Harrison, P.M. Latter & D.W.H. Walton, 155-165. (ITE symposium no. 24.) Grange-over-Sands: Institute of Terrestrial Ecology.

Latter, P.M. & Howson, G. 1977. The use of cotton strips to indicate cellulose decomposition in the field. *Pedobiologia,* **17,** 145-155.

Malmer, N. 1969. Organic matter and cycling of nutrients in virgin and present ecosystems. *Oikos* (suppl.), **12,** 79-86.

Miller, H.G., Cooper, J.M., Miller, J.D. & Pauline, O.J.L. 1979. Nutrient cycles in pine and their adaptation to poor soils. *Can. J. For. Res.,* **9,** 19-26.

Nys, C. & Howson, G. 1988. Effects of tree species on forest soils in northern France, detected by cotton strip assay. In: *Cotton strip assay: an index of decomposition in soils,* edited by A.F. Harrison, P.M. Latter & D.W.H. Walton, 90-93. (ITE symposium no. 24.) Grange-over-Sands: Institute of Terrestrial Ecology.

Olsen, J.S. 1963. Energy storage and the balance of producers and decomposers in ecological systems. *Ecology,* **44,** 322-331.

Piene, H. & Van Cleve, K. 1978. Weight loss of litter and cellulose bags in a thinned white spruce forest in interior Alaska. *Can. J. For. Res.,* **8,** 42-46.

Snedecor, G.W. & Cochrane, W.G. 1967. *Statistical methods.* 6th ed. Ames, Iowa: The Iowa State University Press.

Turner, J. & Long, J.N. 1975. Accumulation of organic matter in a series of Douglas-fir stands. *Can. J. For. Res.,* **5,** 681-690.

Whittaker, R.H. 1975. *Communities and ecosystems.* New York: Macmillan.

Witkamp, M. & Ausmus, B.S. 1976. Processes in decomposition and nutrient transfer in forest ecosystems. In: *The role of terrestrial and aquatic organisms in decomposition processes,* edited by A. Macfadyen, 375-396. Oxford: Blackwell Scientific.

Demonstrating effects of clearfelling in forestry and the influence of temperature and moisture on changes in cellulose decomposition

P M LATTER and F J SHAW
Institute of Terrestrial Ecology, Merlewood Research Station, Grange-over-Sands

Poster summary
When a forest is felled, the alteration of conditions affecting decomposition are complex. Litter input is reduced, the microclimate is altered and the various effects of living roots are replaced by those of dead decomposing roots.

The cotton strip assay was used to assess the effects of felling on decomposition processes in the soil profile of felled and unfelled plots at Kershope Forest, Cumbria, during 2 years following felling.

Cellulose decomposition increased after felling, to a depth of 20 cm (the lowest depth to which strips were inserted), and the higher decomposition was associated with higher litter humidity and generally higher surface temperature on the felled plots. Humidity in the litter region was recorded intermittently with a Vaisala humidity probe, and the thermal cell method (Ambrose 1980) was used to obtain an integrated measure of soil temperature during the period of strip insertion.

The decomposition rate at Kershope was lower than those recorded on similar plots in north Wales at Beddgelert and in north Devon (Torridge).

The work is part of a larger study examining other effects of clearfelling at a number of sites, and will be fully reported elsewhere.

Reference
Ambrose, W.R. 1980. Monitoring long-term temperature and humidity. *Bull. Inst. Conserv. Cult. Mater.*, **6,** 36-42.

Patterns of decomposition assessed by the use of litter bags and cotton strip assay on fertilized and unfertilized heather moor in Scotland

D D FRENCH
Institute of Terrestrial Ecology, Banchory Research Station, Banchory

1 Summary

Tensile strength losses from cotton strips (CTSL), weight losses from 2 plant litters, and the responses of cotton and litter decay to soil amelioration were compared, using data from treated and untreated plots on a Scottish moor. The rate of decomposition of cotton strips, and the magnitude of responses to nutrients or carbohydrates added to the soil were generally intermediate between decomposition of heather (*Calluna vulgaris*) stems and purple moor-grass (*Molinia caerulea*) leaves, but the detailed patterns (shapes) of their responses to additives were very different.

Some wider comparisons are referred to, and it is generally is concluded that CTSL is a satisfactory index of general decomposer activity in the field, at least in cool temperate and sub-polar sites, if comparisons between sites or plots are made on a fairly gross scale (ie involving appreciable differences in climatic or edaphic parameters) or if only a rank order comparison is required. Cotton strip assay should not be used to estimate absolute decay rates or responses to environmental changes of natural litters, or as a means of comparing very different assemblages of substrates. However, the assay can, in the latter case, help to distinguish environmental from substrate 'quality' effects on decomposition rates.

2 Introduction

Much of the justification for the use of the cotton strip assay in ecological studies lies in the 2 related ideas that:

i. cotton strip tensile strength loss (CTSL) is a good index of cellulose decomposition rates in an ecosystem;
ii. cellulose decomposition rates can be used as a general index of the processes of organic matter decomposition and nutrient release.

Howard (1988) is strongly critical of both these assumptions, but direct experimental evidence appears to be scanty.

An experiment to test the effects of added nutrients or carbon sources on decomposition rates on a Scottish moor provided an opportunity, albeit limited, to compare the patterns of decay shown by CTSL with weight losses from plant litters (in bags), using 2 contrasting litters: purple moor-grass leaves and heather stems.

The following questions, related to the 2 assumptions

above, were addressed. How does CTSL compare with weight losses from plant litters? Is the response of CTSL to changes in soil conditions comparable with the responses of weight losses from litters? Do the results suggest that CTSL is a good index of overall rates and processes of decomposition and nutrient release under field conditions?

3 Site description

The study site was a very uniform heather moor at Glen Dye, Kincardineshire. (Miles (1973) gives a general site description, and some climatic data are given by French (1988a).) The experimental area was mown during the spring of 1979 and the mowings removed. Neither the litter layer nor the underlying soil was disturbed or visibly altered by the mowing. Litters and cotton were inserted in late October of that year, after treatment of the experimental plots. (For details of experimental layout, see French (1988c).)

The plot treatments were:

i. untreated controls

ii. N (ammonium nitrate) at 4, 16 and 40 g m^{-2}

iii. P (potassium dihydrogen orthophosphate) at 6, 24 and 60 g m^{-2}

iv. Ca (calcium carbonate) at 10, 40 and 100 g m^{-2}

v. CHO (carbohydrate–mixture of glucose and potato starch 1:1) at 4, 16 and 40 g m^{-2}

The 3 levels of each additive are hereafter referred to as levels 1, 4 and 10, being those multiples of the lowest levels applied (thus N 1, P 4, etc). All additives were evenly broadcast in dry form, at the above levels, on 3 occasions during September and October 1979, before the cotton and litters were put out. There were then no further additions during the experiment.

4 Materials and methods

Shirley Soil Burial Test Fabric (Walton & Allsopp 1977), in 30 cm x 10 cm strips, was wrapped around a turf cut in the upper soil and litter, following the procedure of French and Howson (1982). Of the 5 substrips used in this method, I used data from the top 2 lying horizontally in the litter layer, these being most directly comparable with the decomposing litter in litter bags.

The moor-grass leaves had been gathered immediately after autumnal senescence, so that they were all of almost identical condition. The heather stems had been selected to be between 2 mm and 4 mm in diameter, and cut to 10 cm long. Each litter was dried at 40°C, and 4 g samples were then wrapped in nylon hair nets and placed in the litter layer at the site. Litter bags were retrieved only at 45 weeks. Ingrowing mosses and lichens, extraneous litter fragments, and animals were removed (but not invertebrate faeces), and each sample was dried at 40°C and weighed to determine the weight loss from the sample.

Cotton strips were inserted and retrieved in 3 consecutive batches at 0–19, 19–34 and 34–45 weeks. The test substrips (frayed to 3 cm width) were prepared according to Latter and Howson (1977) and tensile strength (TS) was measured on a Monsanto tensometer. Cloth controls and field controls in all plots and sample periods had identical mean TS values, so all final TS measurements could be related to a single original figure when calculating CTSL.

CTSL was corrected for cementation (French 1984, 1988b), and litter weight losses for ground contact, by linear regression (French 1988c), before analysing the results.

5 Results
5.1 Annual weight losses
An estimate of the total decomposition of cotton strips over the 45-week period may be obtained simply by summing CTSL over all 3 batches (Table 1). The total percentage CTSL ranged from 69% in control plots to 85% on plots receiving the higher levels of Ca application. The data in Heal et al. (1974) and Latter et al. (1988) were used to estimate a relationship between CTSL and weight loss of cotton strips, and the observed CTSL was converted to weight loss. The total CTSL represented weight losses of cotton of about 15–18%. This estimate involves some replacement of fresh material as it decays. An alternative estimate, calculating the expected total weight loss if the first batch of cotton had been left for 45 weeks, assuming a negative exponential decay curve modified only by seasonal effects, gives expected weight loss of about 30–50%. The corresponding range of weight losses over 45 weeks for heather stems was 3–8%, and for moor-grass leaves 30–40%, so the observed CTSL converts to a range of weight losses either intermediate between results for the 2 litters or close to the weight losses from moor-grass leaves.

5.2 Changes in decomposition rates by soil treatments
The overall response of CTSL to the 4 soil additives is again intermediate between the 2 litters (Table 1), with a mean increase over untreated plots of 14%, compared to 9% for moor-grass or 111% for heather. Like the second (exponential) estimate of annual weight loss rates (Section 5.1), the size of changes in CTSL with added nutrients or carbon is more like moor-grass (whose range it overlaps considerably) than heather. The significance of differences from

Table 1. Tensile strength losses (CTSL) from cotton strips, and weight losses (%) from plant litters, after 45 weeks in fertilized and unfertilized plots, with percentage difference from controls

	Heather stems				Cotton strips				Moor-grass leaves			
		Weight loss (%)		% difference from control		CTSL (kg)		% difference		Weight loss (%)		% difference from control
Treatment	n	Mean	SE		n	Mean	SE	ence	n	Mean	SE	
Control (untreated)	19	2.4	0.37		20	116	3.5		18	30.4	0.65	
N 1	20	4.5	0.27	88*	20	134	2.4	16*	18	30.7	0.78	1
N 4	20	3.9	0.28	63*	20	129	2.6	11*	18	32.3	0.84	6*
N 10	19	3.7	0.36	54*	20	130	2.4	12*	18	32.6	0.49	7*
P 1	19	7.7	0.25	221*	20	128	1.9	10*	(10	40.2	1.52	32)[1]
P 4	19	6.8	0.17	183*	20	135	2.1	16*	18	30.1	1.45	−1
P 10	19	8.0	0.18	233*	20	121	2.8	4	18	31.8	1.06	5
Ca 1	18	4.2	0.36	75*	20	131	2.2	13*	17	33.8	1.27	11*
Ca 4	20	3.3	0.15	38*	20	145	2.9	25*	17	33.0	1.45	9
Ca 10	20	3.4	0.23	42*	20	142	1.9	22*	17	36.6	0.71	30*
CHO 1	20	6.0	0.33	150*	20	134	2.0	16*	18	35.8	0.56	18*
CHO 4	20	4.6	0.18	92*	20	134	2.4	16*	18	34.2	0.58	13*
CHO 10	19	4.6	0.23	92*	20	121	3.2	4	18	29.1	0.76	−4

N, nitrogen; P, phosphorus; Ca, calcium; CHO, carbohydrate. 1, 4, 10 are relative amounts (proportion of lowest level) of additive in each treatment

[1] Many moor-grass bags in this treatment were chewed by voles. All obviously chewed bags were rejected, but the weight loss from the remainder may still be higher than the actual decomposition loss

* $P < 0.05$ (Mann–Whitney U test)

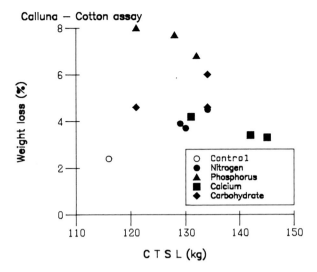

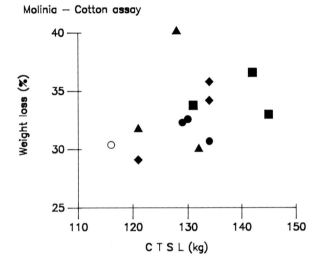

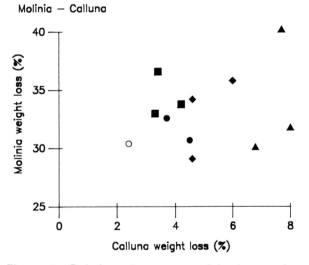

Figure 1. Relations between weight losses from heather (Calluna) and moor-grass (Molinia) litters and tensile strength loss from cotton strips (CTSL) on fertilized and unfertilized plots at Glen Dye

control plots follows a similar pattern. All treatment differences in heather are significant (P<0.05), all but 2 in cotton, but only 6 are significant in moor-grass (Table 1).

As well as the size of response to increased soil

Table 2. Correlations between mean decay rates (per treatment) of cotton strips and plant litters

	All treatments	Omitting P 1	Omitting all P treatments
Heather/cotton	−0.15	−0.12	0.15
Cotton/moor-grass	0.35	0.58*	0.68*
Heather/moor-grass	0.27	−0.11	0.25

* P<0.05 (n = 13)

nutrients and energy sources, the patterns of change may also be examined. To what extent do the 3 substrates respond similarly to the changed soil conditions? If all responses follow essentially the same pattern, then there should be significant correlations between the decay rates of any 2 substrates, over all treatments together. The experimental layout did not allow matching of individual samples, but means for each treatment can be correlated (Figure 1 & Table 2). The response of heather wood decomposition to all levels of P was exceptionally high, and the large increase in losses from moor-grass in treatment P 1 might be spurious (due to removal of material by voles (Muridae), see note to Table 1). Therefore, the correlations in Table 2 are calculated (i) on all points, (ii) excluding treatment P 1, and (iii) excluding all P treatments. Only cotton and moor-grass show a significant relationship in any of these analyses; even the highest correlation between them accounts for less than half of the total variance, and the 2 significant correlations both exclude treatment P 1.

There are, then, no simple linear relationships between heather and either of the other substrates over all treatments, with or without P, but there may be a broad correlation between cotton and moor-grass. However, if the responses of individual substrates to any one additive were to follow some kind of nonlinear, possibly 'humped-shaped' curve, as described by Heal *et al.* (1981, pp619–620) for response to temperature and moisture, a search for simple linear correlations may be unjustified. Are there any other consistent patterns of responses to treatments, eg a general ranking of optimal levels of each additive among the 3 materials? In such a case, the responses could all follow the same pattern, but show no linear correlation between substrates.

In Figure 2, all responses are expressed as a percentage of the maximum response to each additive by each substrate, so that all are on a common scale, emphasizing the patterns, rather than magnitude, of the responses. There are not sufficient data to calculate true response curves, but the simplest probable ordering of responses was estimated by Jonckheere's (1954) S test on the original weight loss or CTSL for each substrate and additive, and these orders are indicated by the lines between points in Figure 2.

The level of each additive producing the largest increase in decay rates of heather wood was generally

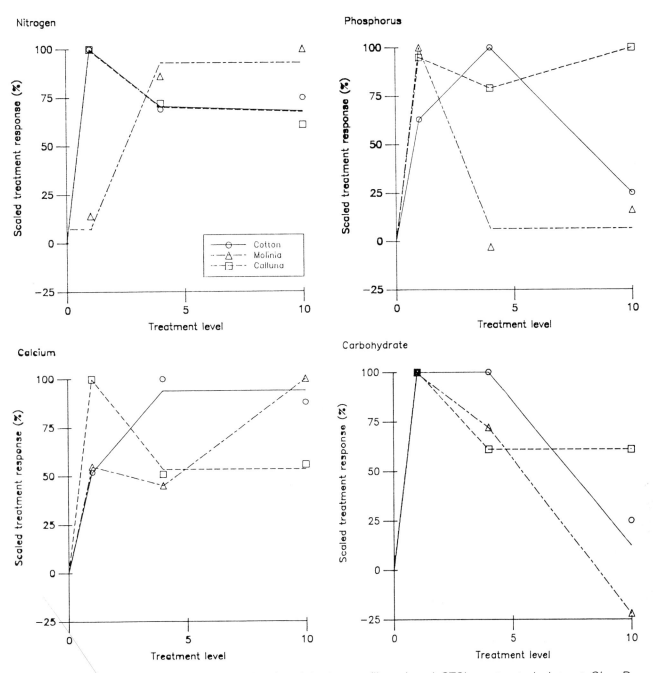

Figure 2. Mean differences from controls of weight losses (litters) and CTSL on treated plots at Glen Dye, scaled to a common range to compare shapes of responses. Lines indicate ordering of responses by Jonckheere S tests. A horizontal line joining 2 treatment levels indicates no significant difference in response between those 2 levels (eg cotton, Ca 4 to Ca 10; all substrates, N 4 to N 10). Similarly, the right-hand side of the cotton/carbohydrate line indicates no difference between CHO 10 and controls

level 1, and of cotton level 4. Moor-grass leaves had either not reached an 'optimal' level by level 10, or had a peak near level 1 followed by a more rapid decline than with either of the other 2 substrates. Optimal levels for cotton, then, are intermediate between the 2 litters, like nearly all other characteristics measured so far.

If, however, the actual shapes of the responses (as defined by the Jonckheere S tests) are compared, it is clear that in no case do all 3 substrates show the same pattern of response. In particular, 3 of the 4 sets of responses for cotton, to P, Ca, CHO additions, may

be interpreted as simple 'humps', rising to a maximum response and then declining at higher levels. Responses by moor-grass include one hump (with CHO), 2 more complex patterns, probably including at least 2 turning points (with added P and Ca), and a response to N that might be the beginning of either (or neither). All 4 responses of heather are very similar to each other, and 3 are very different to both the other substrates, though its response to added N is remarkably close to that of cotton.

6 *Discussion*
In having only 2 litter types, the data from this study

Plate 6. Insertion of cotton strip in the field (Photograph G Howson)

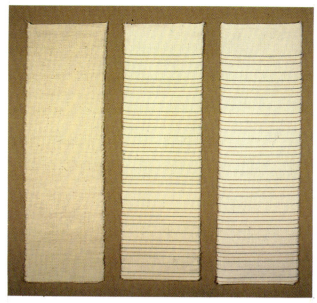

Plate 7. The original unbleached calico cloth and the Shirley Soil Burial Test Fabric (1976 green, and 1981 blue) used for cotton strip assay (Photograph P A Coward)

Brown earth   Peaty gley   Podzol

Plate 8. Pigmentation of cotton produced after growth of fungi with 3 soils:
 i. Chrysosporium pannorum   ii. Chrysosporium merdarium
(Photographs J Gillespie)

Plate 9. Mississippi floodplain mixed hardwood sites used for cotton strip assay
 i. flooded   ii. non-flooded
(Photographs E Maltby)

can only provide a limited comparison between results from the cotton strip assay and from litter decomposition as measured by weight loss. However, in the absence of any other comparable information obtained by simultaneous use of the 2 methods, it is important to make this preliminary examination of any relationships between the 2 measures of decomposition processes in order to highlight some of the problems involved. Comparison of the 2 measures on a seasonal basis is also discussed in French (1988a).

6.1 Substrate quality and decomposition

Using similar criteria to Heal and French (1974), the 3 substrates can be ranked in order of overall 'quality', and their decay rates compared with their quality rank. Cotton strips are essentially devoid of mineral nutrients, but they contain no lignin or other direct microbial inhibitors, and are not physically or structurally especially difficult to decompose. Moor-grass leaves have only moderate lignin content, are low in inhibitory compounds, relatively high in mineral nutrients (1.5–2.0%) and present few physical or structural barriers to decomposers. Their general 'quality' is thus rather higher than that of cotton. Conversely, heather stems have a high lignin content (>25%), a considerable quantity of other inhibitory substances (eg tannins), fairly low nutrient content (total nutrients <1%), are physically hard and structurally intransigent. These are all features of low-quality substrates. The order of substrate quality is therefore: heather<cotton<moor-grass. The decay rates of the 3 substrates are in the same order as their quality ranking, and their overall responses to soil additives, as to be expected, in the reverse order. Their differing patterns of response to different levels of additives may also be related to their differing physical and chemical compositions, and the degree to which these require specialized micro-organisms for their breakdown.

Despite the limitations of the data, it is possible to smooth the responses derived from the Jonckheere tests, taking into account the decomposition mechanisms which are possibly operative, to give a (partly speculative) series of full response curves (ie interpolated and smoothed between actual data points) which are fully consistent with the available data (Figure 3). The one case where the response of cotton is not a simple hump (indicating an optimal level for the full microflora decomposing a single simple substrate) is with addition of N. Widden et al. (1986) have shown that there is selection of microflora by some of my soil treatments, and also selection by cotton strips for particular cellulolytic organisms. My own field observations (French 1988c) suggest selection of different groups of organisms on cotton at different levels of added N, but not of P or Ca, where it seems that only the intensity and not the direction of microbial selection varies between treatment levels. The responses of cotton strips to added N within the 3 sample periods (which were summed for comparison

with litter weight losses) form a series from an immediate response to low levels of N which quickly declines, to little or no initial response to high levels, but the response increasing with time (French 1988c). This pattern suggests that low levels of N increase the activity of the native microflora, without significant selection, the increase ceasing as the added N is used up. However, higher levels of N, in combination with the presence of cotton cellulose, alter the microflora through selection for a group which can efficiently use both cellulose and the extra N. This subgroup, however, takes some time to build up in the overall population, so it shows no effect for several months after N addition. Combining these 2 patterns gives a curve of the type shown in Figure 3.

Heather will select strongly for ligninolytic organisms, so mechanisms of decomposition are likely to be similar. Selection by substrate could conflict with selection by treatment, as many lignin decomposers are intolerant of, or do not compete well in, high nutrient conditions, while others, including many basidiomycetes, have particular requirements for combinations of nutrients (and other environmental conditions). Hence, at any one treatment level, there is likely to be only a limited proportion of the total microflora able to decompose heather wood under those conditions. The strongest influence on decay rates here is likely to be the physical and chemical intransigence of the heather wood, and it is notable that all responses of heather in Figure 3 follow essentially the same pattern. This is not the case with the other 2 substrates; with them, it is practically impossible to fit a single shape of curve to the data for all 4 additives.

Moor-grass leaves, with a more balanced mixture of constituents than either heather or cotton, would not be expected to show any simple limits, or exert any major microbial selection pressure, and generally this substrate shows a more complex array of responses than heather or cotton.

The detailed relationships between substrate quality and patterns of decomposition are discussed further elsewhere (French 1988c). Perhaps more important in the context of this paper is the distinction between the simple 'humps' characteristic of the responses of cotton (a simple, almost pure, single substrate) and the more complex or varied responses of the 2 litters (which both contain a mixture of substrates). Yet again, however, the cotton is intermediate, this time in its degree of variation, between the complete consistency of heather's responses and the very varied responses of moor-grass. In many ways, heather wood is a more distinctive substrate (cf also the correlations between decay rates, Table 2) and, generally, the 2 litters are no more similar than either of them and cotton.

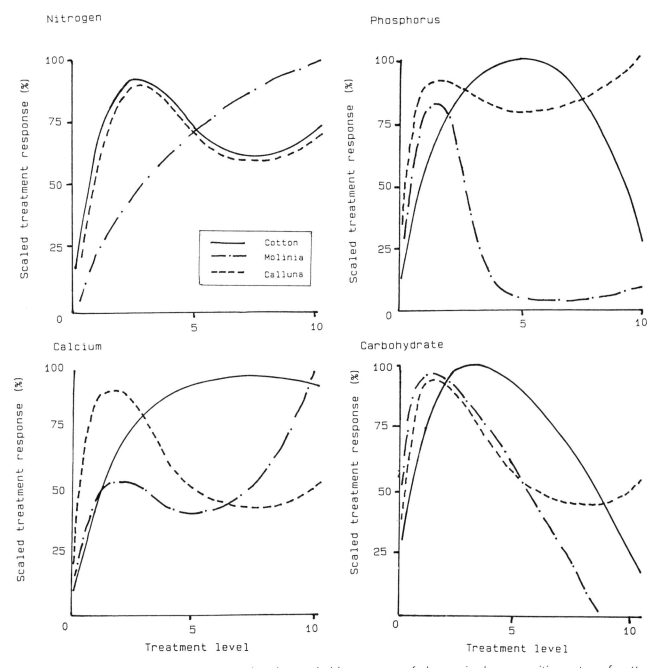

Figure 3. Smoothed response curves, showing probable patterns of change in decomposition rates of cotton and of plant litters with soil additives, derived from the data in Figure 2

6.2 International Biological Programme (IBP) decomposition studies – a wider comparison

The results presented here suggest that, on average, cotton strips can be used as a broad general comparative index of decomposer activity (eg in relation to environmental factors), but that detailed extrapolation to the behaviour of assemblages of more 'natural' substrates may not be possible. This qualification, from my results, also applies to the use of any single substrate as a general indicator of the behaviour of others. In the IBP tundra biome studies (Heal *et al.* 1974), both natural litters and cotton strips were used in a wide range of sites (albeit not always in strictly comparable situations), and the relation between their decay rates and site characteristics can be compared.

Both cotton and natural litters showed similar responses to temperature and moisture (Heal & French 1974; Heal *et al.* 1974, 1981), except at very high moisture levels, where decomposition of litters often continued after cotton strip decay had ceased. The response surface was also flatter with litters, partly because of the greater variation due to substrate quality. Responses to edaphic conditions were less comparable, but CTSL and litter weight losses showed very similar trends when superimposed on a principal component analysis of climate and soil variables. From these studies, then, there appears to be a general agreement of results derived from CTSL and litter bag weight losses, when they are compared on a broad scale. The experiments discussed here for Scottish moorland indicate the probable limits of that agreement when the scale of comparison is narrower.

7 The use of the cotton strip assay

At Gisburn Forest, tree growth, nutrient release and nutrient uptake by trees were all related to CTSL in soils (Brown 1988). On a broader scale, the IBP studies discussed above showed strong parallels between decomposition of cotton and plant litters in their response to environmental variation. Other studies have given similar results, involving inter-relationships between cotton strips, other decomposition measurements, and nutrient release and turnover. Yet Howard (1988) concludes that the breakdown of pure cellulose added to soil cannot provide an index of litter decomposition rate, release of litter nutrients, or 'general biological activity'. This apparently contradictory conclusion is justified by his reference to poor conceptual models of 'soil physiological systems', and an argument from a particular definition of cellulose decomposition through a *reductio ad absurdum* of the chain of connections from CTSL through cellulose decomposition, litter decay and soil organic matter decomposition, to nutrient release. Howard's own arguments are themselves open to criticism; for example, he ignores the many alternative chains connecting CTSL to nutrient release, such as CTSL being a measure of physical penetration or 'opening up' of the substrate, which may be the process most limiting access to, and hence release of, nutrients and other resources contained therein, with no need to invoke carbon loss rates in any direct form. Nevertheless, his critical examination of the models (actual or implied) involved in much of the use of cotton strips should be considered carefully by those using the method. This consideration is especially important if the cotton strip assay is used as an indirect index of other processes, when the kind of model assumed, in relation to the purpose of the measurement, may be critical to interpretation of results.

8 Acknowledgements

I thank Sir William Gladstone, Bt, for permission to work at Glen Dye. Many people assisted with field or laboratory work, or provided useful advice and criticism during the study.

9 References

Brown, A.H.F. 1988. Discrimination between the effects on soils of 4 tree species in pure and mixed stands using cotton strip assay. In: *Cotton strip assay: an index of decomposition in soils*, edited by A.F. Harrison, P.M. Latter & D.W.H. Walton, 80-85. (ITE symposium no. 24.) Grange-over-Sands: Institute of Terrestrial Ecology.

French, D.D. 1984. The problem of 'cementation' when using cotton strips as a measure of cellulose decay in soils. *Int. Biodeterior.*, **20**, 169-172.

French, D.D. 1988a. Seasonal patterns in cotton strip decomposition in soils. In: *Cotton strip assay: an index of decomposition in soils*, edited by A.F. Harrison, P.M. Latter & D.W.H. Walton, 46-49. (ITE symposium no. 24.) Grange-over-Sands: Institute of Terrestrial Ecology.

French, D.D. 1988b. The problem of cementation. In: *Cotton strip assay: an index of decomposition in soils*, edited by A.F. Harrison, P.M. Latter & D.W.H. Walton, 32-33. (ITE symposium no. 24.) Grange-over-Sands: Institute of Terrestrial Ecology.

French, D.D. 1988c. Some effects of changing soil chemistry on decomposition of plant litters and cellulose on a Scottish moor. *Oecologia.* In press.

French, D.D. & Howson, G. 1982. Cellulose decay rates measured by a modified cotton strip method. *Soil Biol. Biochem.*, **14**, 311-312.

Heal, O.W. & French, D.D. 1974. Decomposition of organic matter in tundra. In: *Soil organisms and decomposition in tundra*, edited by A.J. Holding, O.W. Heal, S.F. MacLean & P.W. Flanagan, 279-310. Stockholm: Tundra Biome Steering Committee.

Heal, O.W., Howson, G., French, D.D. & Jeffers, J.N.R. 1974. Decomposition of cotton strips in tundra. In: *Soil organisms and decomposition in tundra*, edited by A.J. Holding, O.W. Heal, S.F. MacLean & P.W. Flanagan, 341-362. Stockholm: Tundra Biome Steering Committee.

Heal, O.W., Flanagan, P.W., French, D.D. & MacLean, S.F. Jr. 1981. Decomposition and accumulation of organic matter. In: *Tundra ecosystems: a comparative analysis*, edited by L.C. Bliss, O.W. Heal & J.J. Moore, 587-634. Cambridge: Cambridge University Press.

Howard, P.J.A. 1988. A critical evaluation of the cotton strip assay. In: *Cotton strip assay: an index of decomposition in soils*, edited by A.F. Harrison, P.M. Latter & D.W.H. Walton, 34-42. (ITE symposium no. 24.) Grange-over-Sands: Institute of Terrestrial Ecology.

Jonckheere, A.R. 1954. A distribution-free k-sample test against ordered alternatives. *Biometrika*, **41**, 33-145.

Latter, P.M. & Howson, G. 1978. The use of cotton strips to indicate cellulose decomposition in the field. *Pedobiologia*, **17**, 145-155.

Latter, P.M., Bancroft, G. & Gillespie, J. 1988. Technical aspects of the cotton strip assay method. *Int. Biodeterior.*, **24**. In press.

Miles J. 1973. Natural recolonization of experimentally bared soil in Callunetum in north-east Scotland. *J. Ecol.*, **61**, 399-412.

Walton, D.W.H. & Allsopp, D. 1977. A new test cloth for soil burial trials and other studies on cellulose decomposition. *Int. Biodeterior. Bull.*, **13**, 112-115.

Widden, P., Howson, G. & French, D.D. 1986. Use of cotton strips to relate fungal community structure to cellulose decomposition rates in the field. *Soil Biol. Biochem.*, **18**, 335-337.

Effects of lime and pasture improvement on cotton strip decomposition in 3 Scottish acid hill soils

P J VICKERY[1] and M J S FLOATE[2]
Hill Farming Research Organisation, Penicuik

1 Summary

In 1978, cotton strips were buried for 36 days in 3 sites under pasture with a control plot and 4 pasture improvement treatments, including lime, basic slag, rye-grass (*Lolium perenne*) and white clover (*Trifolium repens*), applied 6–9 years earlier. All the sites were grazed by wether sheep 3 times per year.

Decomposition of cellulose, measured as loss in tensile strength of cotton strips, was greatest in the 0–4 cm depth layer of all limed treatments, and was least in the deepest (8–12 cm) of the control treatments. Pasture improvements other than lime had little effect on decomposition rate. Decomposition of cellulose was greater in the brown earth soil than in the 2 more acid peaty podzolic soils.

2 Introduction

Improvement of natural and semi-natural grassland by the use of introduced plants and fertilizers usually results in a substantial increase in annual net primary production, and concomitant increases in the rate of nutrient cycling through enhanced decomposition of soil organic matter by invertebrates and micro-organisms (Floate *et al.* 1981; Hutchinson & King 1984). The cotton strip assay was used as an integrated measure of these decomposition processes in experimental plots with a range of pasture improvement treatments on 3 hill country soils.

3 Site and treatment description

Pasture improvement trials were established on 3 contrasting soil types at the Hill Farming Research Organisation's Field Research Station, Sourhope, Roxburghshire, UK, in 1969, 1970 and 1971. The soils ranged from a brown earth (pH 5.0–5.5), through thin peaty podzol (pH 4.0–4.5), to a thick peaty podzol (pH <4.0), and were situated at 300, 350 and 370 m above sea level (55.5°N, 2.2°W). Semi-natural grassland

swards on these soils consisted of bent-grass/fescue (*Agrostis/Festuca*), mat-grass/fescue (*Nardus/Festuca*), and moor-grass/mat-grass (*Molinia/Nardus*) respectively. Additional soil details are given in Table 1.

The pasture improvement treatments were applied to the plots as follows:

C control;
  i. lime (at 6.3 t ha^{-1});
 ii. lime + basic slag (at 1.25 t ha^{-1});
iii. lime + basic slag + white clover (at 2.2 kg ha^{-1});
 iv. lime + basic slag + white clover + rye-grass (at 22 kg ha^{-1}) with 40 kg^{-1} ha nitrogen (N).

The plots were grazed by wether sheep for 3 30-day periods (starting in mid-May, mid-July and mid-October), at stocking rates adjusted to dry matter present at the start of each period, so that 80–90% of the herbage was consumed in the 30-day grazing period. Full details of treatments and herbage production are given in Eadie *et al.* (1981) and Newbould (1981). Over the 10 years of the trials, annual pasture production from the control sites ranged between 1.5 and 2.5 t ha^{-1}, while that from the fully improved sites ranged between 4.0 and 6.0 t ha^{-1}.

4 Experimental details

Using the standard technique (Latter & Howson 1977), cotton strips were inserted in the soil profiles at the 3 experimental sites on 26 June 1978 (summer) and removed 36 days later. The strips were inserted in pairs at 5 locations within each pasture improvement treatment. Depth of insertion of the strips varied between soils due to stones and compactness of the lower soil horizons; the maximum depth of insertion possible was 25 cm, but this was rarely achieved; the actual depths are given in the results. At the time of insertion, 2 strips from one location per treatment

Table 1. Pre-treatment soil characteristics of soil types used in Sourhope pasture improvement experiment

Soil characteristic	Soil types studied		
Type	Acid brown podzolic (brown earth)	Peaty podzolic (thin)	Peaty podzolic (thick)
Approximate profile depth (cm)	35	10–20	35–40
Surface texture	Silty loam	Silty loam	Humic silty loam
Drainage	Free	Poor	Impeded and variable
Depth A$_o$ organic layer (cm)	2.0	5–10.0	10.0
pH (0–5 cm)	5.0–5.5	4.0–4.5	4.0
Dominant vegetation	Bent-grass/fescue	Mat-grass/fescue	Moor-grass/mat-grass

[1] on sabbatical leave from *CSIRO, Pastoral Research Laboratory, Division of Animal Production, Armidale, Australia*
[2] now at *Ministry of Agriculture and Fisheries, New Zealand*

were removed for field controls, leaving the overall experimental design as follows:

3 soil types x 5 pasture improvement treatments x 4 locations within treatments x 2 replicates within locations

At each location, the pairs of strips were covered with a small wire mesh cage to prevent disturbance by sheep grazing the plots. After removal, the cotton strips were taken to the laboratory, carefully washed, divided into 4 cm substrips and each frayed back to 3 cm. This represented 4 depths of measurement at 0–4 cm, 4–8 cm, 8–12 cm, and 12–16 cm. The divided strips were then sent to the Institute of Terrestrial Ecology's Merlewood Research Station for tensile strength (TS) testing. Some temperature and humidity data were recorded at 2 sites (Table 2).

Table 2. Climatic information for the site
i. Environmental conditions during experiment (1978)

	Temperature (°C)	Relative humidity (%)
Mean maximum	13.6	99%
Mean minimum	6.6	81%

ii. Long-term average environmental conditions of experiment location

	Annual	July
Precipitation (mm)	870	85
Sunshine (h)	1400	160

Statistical analysis of the data for loss in tensile strength of cotton (CTSL), expressed as a percentage:

$$\frac{(\text{field control TS} - \text{final TS at 36 days})}{\text{field control TS}} \times 100$$

was done on natural and transformed (arcsin) data, using a split-plot model, with soils and improvement treatments as main plots split for 3 depth layers. The transformation corrected for some slight skewness in the data but had little effect on the outcome of the analysis of variance. Effects for locations and pairs within the development treatment plots were not extracted, remaining in the error term of the analysis. Data from the deepest layer (12–16 cm) had numerous missing observations and were not used for this report. In the 3 cm depth analyses, there were 9 missing values out of a total of 360 possible observations. Values for these observations were calculated via the standard routine in GENSTAT analysis of variance (Genstat Manual 1977).

5 Results

Rate of decomposition of the cotton strips was significantly affected by the soil type ($P < 0.01$), being greater on the brown earth than on the peaty podzols. The mean CTSL (%) for the 3 soils were 75.4, 70.0, and 61.6 (pooled standard error $\pm$ 1.9%), decreasing in order of soil pH. Soil type also had a significant effect on insertion depth of the strips ($P < 0.001$), the mean depths for the soils being 17, 13 and 22 cm for the brown earth, and the thin and thick peaty podzolic sites, respectively.

Over the 3 soil types, improvement treatment had a very significant influence on CTSL ($P < 0.001$), being least on the control treatments and much greater in all the limed plots (Table 3). The pooled standard error for unconstrained and independent comparison of effects ($\pm$ 2.46%) indicated that the largest effect was the addition of lime, with little difference between the other improvement treatments. Across all the treatments, liming increased the pH of the soils by 1.0, 0.6, and 0.9 units on the brown earth, and the thin

Table 3. Tensile strength loss of cotton strips (CTSL) placed in control and in improved pasture plots at Sourhope. Summary of interaction effects after 36 days' insertion, based on 351 observations and 9 missing values, with the means for the 3 soil types at 3 depths (cm). Field control TS = 50.5 kg; effective standard error = ±4.48

	Treatments				
	Control	Lime	Lime + basic slag	Lime + basic slag + clover	Lime + basic slag + clover + rye-grass
Mean for 3 soil types	50.8	71.1	71.7	77.5	73.9
Brown earth					
0–4	87.7	94.9	88.9	88.8	96.3
4–8	65.6	85.1	82.4	78.5	83.2
8–12	48.1	68.3	68.4	63.9	76.4
Thin peaty podzol					
0–4	47.9	91.1	86.7	98.1	89.5
4–8	35.7	62.2	77.0	88.2	84.5
8–12	23.1	48.0	60.6	77.0	68.0
Thick peaty podzol					
0–4	52.2	92.0	93.8	97.8	95.5
4–8	51.5	74.4	81.7	88.1	86.5
8–12	47.2	46.1	46.4	65.4	38.2

and thick peaty podzolic soils, respectively. Because of the incomplete factorial nature of the experiment, it was impossible to separate the various treatments further. As expected, decomposition decreased with depth; the mean decomposition for all soil types and improvement treatments at 0–4 cm caused an 87% CTSL, while the CTSL at 12–16 cm was 58%.

Many interactions were significant; soils x depth and soil x depth x treatment at P<0.001, treatment x depth at P<0.05; and treatment x soils at P<0.01. Because of the lack of independence between depths and the difference in the maximum depth of insertion between soils, only the data for the first 3 depths were analysed (Table 3). The interactions show that decomposition rates in all soils, as measured by loss in tensile strength of the cotton, responded to the liming treatment but with differing magnitude. The brown earth site showed a higher initial rate for the control treatment, and decomposition of cellulose occurred further down the profile than with the other soils. Variability for all soils and improvement treatments was within expected limits, and the coefficient of variation was 17.4% (Hill *et al.* 1985).

Decomposition of cellulose occurred throughout the brown earth profile, and was faster for the pasture improvement treatments than for the control at all depths. This pattern also occurred with the thin peaty podzolic soil, but with an additional tendency for accelerated decomposition at 8–12 cm under liming, basic slag, white clover and rye-grass treatments. By contrast, the thick peaty podzolic profile only showed accelerated cellulose decomposition for the 2 upper depths, where all limed treatments resulted in greater CTSL than in the control plot. In the upper 2 depths, the rates for treatments on the thick peaty podzolic soil were higher than for the thin peaty podzolic and generally similar to the brown earth, but in the 8–12 cm region the rates for the thick peaty podzolic were considerably lower than for the brown earth.

6 Discussion

In general, the results for cellulose decomposition, as measured by the cotton strip assay for soil type and depth, are consistent with published results, such as those of Heal *et al.* (1974). The peaty podzolic profiles showed results for differences in decomposition between depths which were consistent with the profile types of Heal *et al.* (1974), for peaty and grassed mineral soils respectively, with the thin peaty podzolic being similar to a mineral soil in this instance. These findings are consistent with the peaty nature of the profiles, where the highly acid conditions would either inhibit decomposition or, at least, confine it to the upper aerobic layers of the profile.

Elsewhere (Floate *et al.* 1981), we have noted a strong correlation between post-liming pH of the soils in this experiment and herbage litter decomposition measured in mesh bags, the correlation for pH *versus* dry matter loss rate being r = 0.9336 (P<0.001). However, with loss in tensile strength in the 0–4 cm depth layer, the correlation with pH was not as strong: r = 0.5123 (P<0.1). Also, the correlation of CTSL (0–4 cm) with litter decomposition was low (r = 0.2789 NS), a situation discussed by French (1988). This result was somewhat unexpected because of the obvious overall effect of liming to increase the rate of decomposition of the cotton strips, and the greater rate of decomposition in the brown forest soil compared with the lower pH peaty soils. In reviewing a considerable body of literature on cellulose decomposition, Siu (1951) concluded that bacteria and actinomycetes show greatest activity in slightly alkaline conditions, while acid conditions favour fungal decomposition. Thus, it is possible that the liming treatments resulted in a gross shift in dominance from fungal to bacterial decomposition. Such a change could account for the overall effect of pH in increasing decomposition of the cotton strips, without the 5 treatment differences in pH being specifically related to loss of tensile strength. Further detailed study of the pH liming effect on cotton strip decomposition will be necessary to uncover the mechanism of this effect.

The results from the cotton strip assay have added weight to our hypothesis that the increased pasture production, resulting from the improvement treatments, was partly a consequence of the effect of lime on microbial activity, and hence nutrient cycling. However, between the pasture improvement treatments, the cotton strips did not show a consistent pattern of differences, except that the maximum losses in tensile strength in the peaty podzolic soils occurred with the lime plus basic slag and white clover treatments. The overall similarity of loss of tensile strength, and hence similarity of microbial activity, between pasture improvement treatments requires further investigation to reconcile it with the long-term observed differences in pasture and animal production between the treatments.

7 Acknowledgements

We wish to acknowledge gratefully the assistance of Miss P M Latter and Mrs G Howson, ITE Merlewood, for helping to insert the cotton strips and to test the tensile strength. This study was undertaken when one of us (P J Vickery) was a visiting scientist at the Hill Farming Research Organisation, and use of its facilities for the work is gratefully acknowledged.

8 References

Eadie, J., Hetherington, R.A., Common, T.G. & Floate, M.J.S. 1981. Long-term responses of grazed hill pasture types to improvement procedures I. Nutrient cycling and soil changes. In: *The effective use of forage and animal resources in the hills and uplands*, edited by J. Frame, 167-168. (Occ. Symp. no. 12.) Edinburgh: British Grassland Society.

French, D.D. 1988. Patterns of decomposition assessed by the use of litter bags and cotton strip assay on fertilized and unfertilized heather moor in Scotland. In: *Cotton strip assay: an index of decomposition in soils*, edited by A.F. Harrison, P.M. Latter & D.W.H. Walton, 100-108. (ITE symposium no. 24.) Grange-over-Sands: Institute of Terrestrial Ecology.

Floate, M.J.S., Hetherington, R.A., Common, T.G. & Ironside, A.D. 1981. Long-term responses of grazed hill pasture types to improvement procedures. II. Nutrient cycling and soil changes. In: *The effective use of forage and animal resources in the hills and uplands,* edited by J. Frame, 147-149. (Occ. Symp. no. 12.) Edinburgh: British Grassland Society.

Genstat Manual. 1977. *GENSTAT, a general statistical program.* Oxford: Numerical Algorithms Group.

Heal, O.W., Howson, G., French, D.D. & Jeffers, J.N.R. 1974. Decomposition of cotton strips in tundra. In: *Soil organisms and decomposition in tundra,* edited by A.J. Holding, O.W. Heal, S.F. MacLean & P.W. Flanagan, 341-362. Stockholm: Tundra Biome Steering Committee.

Hill, M.O., Latter, P.M. & Bancroft, G. 1985. A standard curve for inter-site comparison of cellulose degradation using the cotton strip method. *Can. J. Soil Sci.,* **65,** 609-619.

Hutchinson, K.J. & King, K.L. 1984. Biological aspects of nutrient cycling in grazed pasture. The underground movement. *Proc. Aust. Soc. Anim. Prod.,* **15,** 133-136.

Latter, P.M. Howson, G. 1977. The use of cotton strips to indicate cellulose decomposition in the field. *Pedobiologia,* **157,** 145-155.

Newbould, P. 1981. The potential of indigenous plant resources. In: *The effective use of forage and animal resources in the hills and uplands,* edited by J. Frame, 1-15. (Occ. Symp. no. 12.) Edinburgh: British Grassland Society.

Siu, R.G.H. 1951. *Microbial decomposition of cellulose with special reference to cotton textiles.* New York: Reinhold.

Preliminary investigations into use of cotton strip assay for assessing effects of changing land use on soils in mid-Wales

J B DIXON and P F RANDERSON
Department of Applied Biology, University of Wales Institute of Science and Technology, Llysdinam Field Centre, Llandrindod Wells

Poster summary

Land use change surveys indicate that much recent ploughing and reseeding of hill land have been on the brown earth soils of the slopes of the major upland areas. A series of investigations comparing the ecology and management of indigenous bent-grass/sheep's fescue (*Agrostis/Festuca ovina*) grassland with reseeded rye-grass (*Lolium*) swards on brown earth soils of the Manod Association (Dixon 1987) has included assessments of microbial decomposition rates inferred from cotton strip decomposition. Results were presented showing that decomposition rates at 'improved' sites are, as expected, considerably higher than at unimproved sites, and showed little change down the soil profile. The decomposition rate on the unimproved site decreased markedly down the profile, and it is concluded that little mixing of horizons occurs at unimproved sites. As the decomposition in this experiment reached greater than 90% tensile strength loss of cotton in the improved soils, no proper comparisons between individual improved and reverted sites could be made.

Preliminary results of plot experiments to determine the influence of lime on decomposition rates at an unimproved site showed little effect within 60 days of application, although decomposition on the limed plot was 5% higher than on the unlimed plot between 4–16 cm depth.

Further studies will continue to consolidate these research findings, and will include use of other methods to examine litter decomposition and microbial activity. It is considered that a single parameter such as the cotton strip assay is inadequate as a measure of nutrient mineralization rates in these hill soils.

Reference

Dixon, J.B. 1987. Ecology and management of improved, unimproved and reverted hill grasslands in mid-Wales. In: *Agriculture and conservation in the hills and uplands,* edited by M. Bell & R.G.H. Bunce, 32-37. (ITE symposium no. 23.) Grange-over-Sands: Institute of Terrestrial Ecology.

Examination of the biological activity of the soil under natural conditions

K KUZNIAR
Academy of Agriculture, Krakow, Poland

Poster summary

A systematic study was begun in the Putawy State Agriculture Institute, Poland, in 1938 to determine the biological activity of the soil profile. The results of the study were not published until 1948, because of the war.

From 1938 to 1984, the biological activity of the soil in various natural environments in Poland was determined with the aid of strips of cotton or linen cloth. Cellulose decomposition rates in the soils varied according to their nutrient content and physical properties. Using the rate of cellulose decomposition, it was possible to determine the general level of biological activity in soil for use in practical applications of land use. The method was tested in different seasons, in loam, sand, clay and organic soils, in cultivated fields with different crops with or without mineral or organic fertilization, under forest or grassland, or on the border between cultivated fields and forests, in soils with differing moisture contents, and in marshy or mountainous regions.

The following method was used in the study.

Sterile strips of cloth, 150 cm long x 10 cm wide, were placed, if the soil conditions permitted, in narrow vertical slits in the soil. The holes which occurred were filled with soil from the same layer. In some cases, independent of the cloth strips, 9 cm cloth pieces were placed vertically in the profile on glass plates with 12 cm sides. The decomposition of cellulose in the soil, examined with cloth, was estimated after removing the strips from the soil and drying

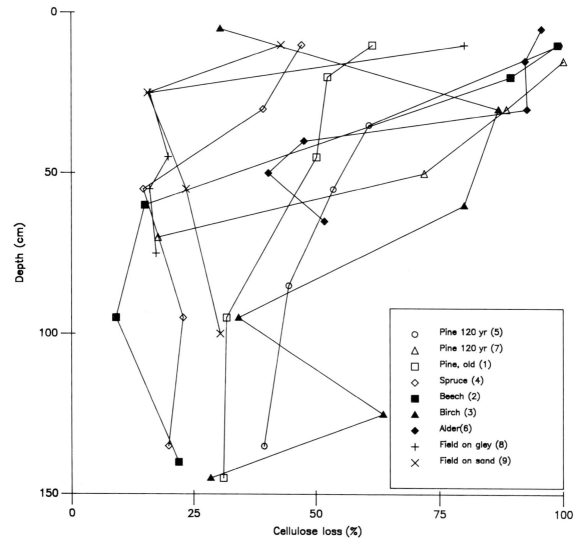

Figure 1. Cellulose decomposition in forest soils in the neighbourhood of Poznan. Site numbers (Table 1) are given in brackets

Table 1. Cellulose decomposition in forest soils in the neighbourhood of Poznan

Site number and location	Forest stand	Soil	Depth (cm)	pH
National Park of Wielkopolska				
1 Puszczykowo Forest	Single old pines	Alluvial soil podzolic with sand	5–15	5.0
			15–25	5.8
			40–50	6.6
			90–100	6.9
			140–150	7.5
2 Puszczykowo Forest	Beech forest 60 years old	Sandy at the top moraine loam, podzolic	5–15	5.1
			15–25	5.3
			55–65	5.4
			90–100	5.8
			35–145	6.2
3 Gorka Forest	Birch forest 20–30 years old	Peat thickly covered with organic silt	0–10	4.2
			25– 35	4.9
			55– 65	4.8
			90–100	4.7
			120–130	4.7
			140–150	
Chief Forestry Ludwikowo				
4 Podloziny Forest	Spruce forest 60 years old	Humus soil with gley on loamy subsoil	5–15	4.8
			25–35	5.7
			50–60	6.7
			90–100	6.9
			130–140	7.2
5 Podloziny Forest	Pine forest 120 years old	Humus soil on loamy subsoil	5–15'	ND
			30–40	ND
			50–60	7.4
			80–90	7.6
			130–145	7.8
6 Wronczyn Rudy Forest	Alder forest 30 years old	Humus soil with gley on clay	0–10	ND
			10–25	6.7
			25–35	7.3
			35–45	7.5
			45–55	7.8
			60–70	7.9
7 Dabrowka Forest	Pine forest 120 years old	Humus brown soil	10–20	6.9
			25–35	7.8
			45–55	7.8
			65–75	7.8
8 Wieckowice Forest	Cultivated field	Humus soil with gley on sandy subsoil	5–15	7.6
			20–30	7.6
			40–50	7.6
			50–60	7.6
			70–80	7.6
9 Dabrowka Forest	Cultivated field	Deep sandy soil	5–15	5.1
			20–30	5.3
			50–60	6.6
			95–105	7.1

ND, not determined

them. The visible changes were described and the degree of decomposition determined using a special scale. Quantitative determination of the cloth decomposition was carried out using analytical methods. The field observations were always made at about the same time.

The results of one study (Kuzniar 1956) are illustrated in Table 1 and Figure 1, and the results of other studies are presented by Kuzniar (1948, 1950, 1952a, b, 1953a, b, c, d, 1956, 1957, 1972, 1976, 1984) and by Skinner and Kuzniar (1975). (The papers are all in Polish, with English summaries.)

References

Kuzniar, K. 1948. Studies on the cellulose decomposition in forest soil. *Inst. Bad. Lesn. Rozpr. i Spraw,* **50,** 1-44.

Kuzniar, K. 1950. New methods of determining a forest soil's activity. *Sylwan,* **3,** 49-57.

Kuzniar, K. 1952a. The energy of cellulose decomposition in the soils of the national park in Pieniny Mountains. *Acta microbiol. pol.,* **1,** 53-58.

Kuzniar, K. 1952b. The energy of cellulose decomposition in the soils of Biatowieza National Park. *Acta microbiol. pol.,* **1,** 257-269.

Kuzniar, K. 1953a. Cellulose decomposition by the microorganisms in forest soil in winter. *Ekol. Polska,* **1,** 137-140.

Kuzniar, K. 1953b. The energy of cellulose decomposition in the soils of the border of cultivated field and forest. *Ekol. Polska,* **1,** 41-53.

Kuzniar, K. 1953c. The influence of the forest border upon the microflora of cultivated soils. *Ekol. Polska,* **1,** 17-39.

Kuzniar, K. 1953d. The influence of the relief upon the biological activity of cultivated loess soils. *Ekol. Polska,* **1,** 31-55.

Kuzniar, K. 1956. The energy of decomposition of cellulose in forest soils. *Ekol. Polska,* **4,** 21-34.

Kuzniar, K. 1957. The influence of various mulches applied in orchards on soil micro-organisms. *Roczn. Nauk roln.,* **76,** 335-367.

Kuzniar, K. 1972. Preliminary results of investigations on the determination of biological activity of garden soils. *Zesz. nauk. wyzsz. Szk. roln. Krakow,* **70,** 201-212.

Kuzniar, K. 1976. Determination of the biological activity of the soil. *Prace Pol. Tow. glebozn.,* 111/19.

Kuzniar, K. 1984. *Introductory information on the methods of investigation used for the determination of the biological activity of soils in natural conditions.* (Bibliography.) Krakow: Academia Rolnicza im. H. Kollataja w Krakowie.

Skinner, F. & Kuzniar, K. 1975. Process of cellulose decomposition in cultivated and forest soils in Rothamsted. *Acad. Rol. w Poznaniu Kom. Mikr. PAN,* 35-37.

Tropical

Use of the cotton strip assay at 3 altitudes on an ultrabasic mountain in Sabah, Malaysia

J PROCTOR[1], G HOWSON[2], W R C MUNRO[1] and F M ROBERTSON[1]
[1]Department of Biological Sciences, University of Stirling, Stirling
[2]Institute of Terrestrial Ecology, Merlewood Research Station, Grange-over-Sands

1 Summary

The cotton strip assay was used in tropical rainforests at 3 altitudes on Gunung Silam (5°N, 119°E), a small coastal mountain (884 m above sea level (asl)) in Sabah, Malaysia. The soils were derived from ultrabasic bedrock (which has low concentrations of phosphorus (P), potassium (K) and calcium (Ca), and high concentrations of magnesium (Mg) and nickel (Ni). The forests on these unusual soils were species-rich, but their stature diminished rapidly with altitude. There was a large variation between replicate cotton strips inserted, particularly at the lowest altitudes, but decomposition was generally rapid with days to 50% loss (CT50) at 0–4 cm depth of 16.2 (280 m), 14.1 (610 m), and 25.0 (870 m). Unexpectedly, the highest rate occurs at the intermediate altitude, suggesting that low nutrient supply is not a cause of the relatively small forest there, but may account, at least partly, for the stunted trees at 870 m.

2 Introduction

Gunung Silam is a small coastal mountain at the eastern end of the Segama Highlands, an area of rugged country built up of igneous, metamorphic and sedimentary rocks. Most of Gunung Silam (including all the plots discussed here) is ultrabasic rock, which supports soils bearing primary rainforest. There is large-stature lowland dipterocarp forest near the base of the mountain, and small non-dipterocarp lower montane forest near the summit. The vegetation shows a classic 'Massenerhebung effect' (Grubb 1971), which refers to the depression of the altitudinal limit of montane forest on small mountains compared with large ones. The physical environment, vegetation and forest processes on Gunung Silam have been described by Proctor *et al.* (1988a, b).

Gunung Silam was visited by Stirling University Expeditions in July–September 1983 and the same

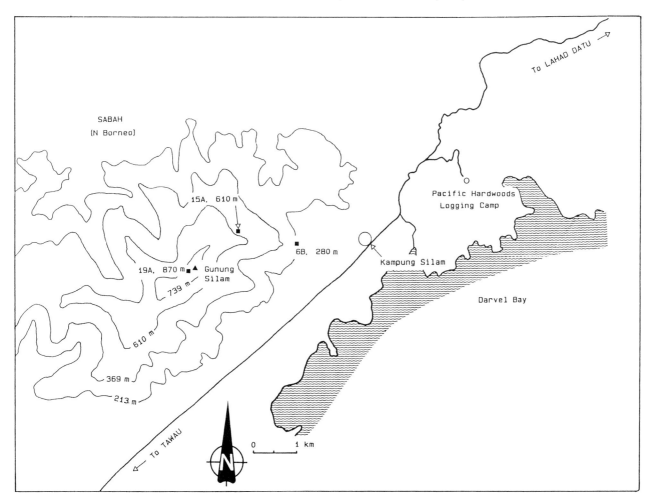

Figure 1. Location of study sites (■)

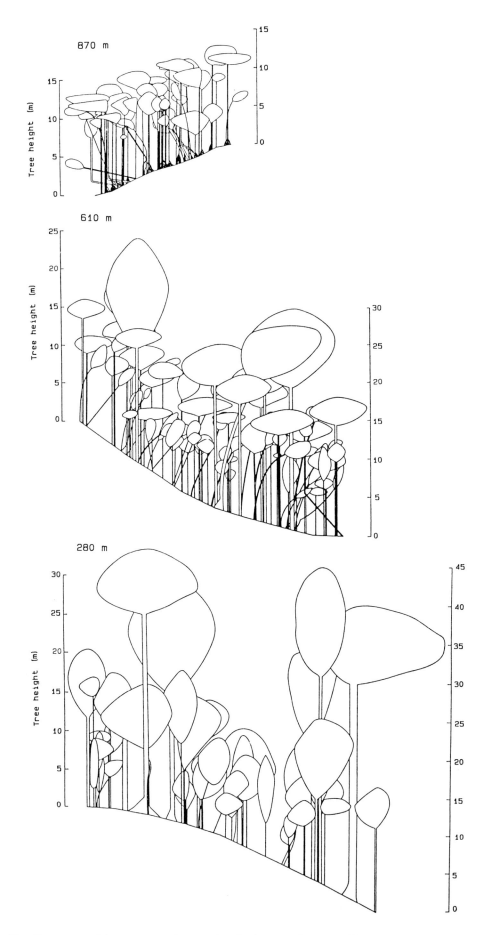

Figure 2. Profile diagrams of forest canopies at the 3 plots at Gunung Silam, Sabah, Malaysia
  i. Plot 6B lowland rainforest at 280 m
 ii. Plot 15A, tall lower montane forest at 610 m
iii. Plot 19A, short lower montane forest at 870 m

Table 1. The Sabah Forest Department plot numbers, the altitudes, the dimensions and a summary of structural and floristic features of 3 rainforest plots on Gunung Silam, Sabah

Sabah Forest Dept plot number	Altitude (m)	Dimensions (m)	Height of highest emergent tree (m)	Height of main canopy (m)	Main families (% basal area)
6B	280	40 × 100	49	30–35	Dipterocarpaceae (27.9) Anacardiaceae (21.5) Verbenaceae (7.8) Leguminosae (7.1)
15A	610	40 × 60	39	20–25	Dipterocarpaceae (27.1) Euphorbiaceae (23.8) Anacardiaceae (12.4) Myrtaceae (10.1)
19A	870	20 × 20	16	10–12	Myrtaceae (60.7) Rutaceae (24.0) Theaceae (5.7) Elaeocarpaceae (2.5)

months in 1984. The influence of the ultrabasic rock on the structure and species composition of the forest was investigated, along with the causes of the 'Massenerhebung effect'. This paper reports our application of the cotton strip assay to test the hypothesis that decomposition rates decrease with altitude, even on small mountains. Grubb (1977) had suggested that low supply rates of nutrients (which would be likely to be directly linked with decomposition rates) might partly account for the occurrence at low altitudes of small-stature vegetation on wet tropical mountains.

3 The study plots

Three plots were selected from several established on the main east ridge of Gunung Silam by the Sabah Forest Department: plot 6B, lowland rainforest at 280 m asl; plot 15A, tall lower montane forest at 610 m asl; and plot 19A, short lower montane forest at 870 m asl (Figures 1 & 2). A summary of the plot features is given in Table 1. Unlike the often stunted or sparse vegetation on ultrabasic soils (Proctor & Woodell 1975), there is no barrenness, and the stunting of the forests near the summit seems part of a general montane effect that occurs on a range of substrata, including ultrabasics.

4 Climate and soils

The mean annual rainfall at a site at 20 m asl about 5 km from the base of Gunung Silam is 2011 mm. There is no regularly defined dry season, although the period June–September (which encompasses the assay dates) is relatively dry. The rainfall for the 1984 assay period for a station at 10 m altitude (5 km from the base of Gunung Silam) and for another at 884 m (on the summit) was 159 mm and 91 mm respectively. Above 650 m, the dryness is offset to some extent by the presence of a frequent cloud cap. Temperature data for the assay period are summarized in Table 2, and are probably representative of the whole year. Fox (1978) has commented that average daily temperatures in Sabah show little variation through the year.

Table 2. The mean screen maximum and minimum temperatures and the mean diurnal range for sites in the open at 10 m and 887 m on Gunung Silam, Sabah. Ranges are given in brackets. The data are for 24 July–12 September 1984

Altitude (m)	Mean maximum (°C)	Mean minimum (°C)	Mean diurnal range (°C)
10	31.7 (27.1–33.9)	23.4 (21.6–25.1)	8.4 (6.2–10.9)
884	27.7 (22.3–31.5)	18.8 (17.3–20.5)	8.9 (3.1–14.0)

The soils on Gunung Silam are brown to yellow-brown, with darker-coloured surface horizons. They are freely drained and have a high proportion of sand (ie particles <0.063 mm), which, in the 0–15 cm depth samples, ranges from about 30% of the dry weight at 280 m to about 25% of the dry weight at 610 m and 870 m (excluding the mor layer). The soils are stony and the average depth (measured at 50 randomly selected points in each plot) to the first stones is 32.6 cm (280 m), 6.1 cm (610 m) and 30.8 cm (870 m). However, sharp stakes could be driven down to 1.8 m at a soil pit in all the plots. At 280 m, the upper horizons had a well-developed crumb structure with a mull humus form and frequent earthworm (Lumbricidae) casts. The forest at 610 m had a weaker crumb structure and more stratification of organic matter in the upper soil horizons. Exposed rock and scree are prevalent at this plot, and the soil there is very stony. At 870 m, there is a brown mor humus (not peat) which is well drained and loosely packed, with extensive fine-root networks which often reach substantial depths (up to one m over prop roots).

The chemical analyses (Table 3) show several important features, including many which are typical of ultrabasic soils (Proctor & Woodell 1975). In general, the samples were mildly acid (pH range 5.3–6.0), but the pH was much less in the mor humus at 870 m (mean 4.0). There was a marked excess of Mg at

Table 3. The means ± SE of pH; loss-on-ignition; exchangeable potassium, sodium, calcium, magnesium; acetic-acid extractable phosphorus, cobalt and chromium; exchangeable nickel; and cation exchange capacity (CEC) in 15 cm deep samples collected from each of 3 plots on Gunung Silam, Sabah. Values are expressed on an oven dried (105°C) basis where appropriate (n = 10 for acetic acid extractable elements and CEC; for other analyses, n = 18 for 870 m and n = 20 for 280 m and 610 m)

Altitude (m)	pH[1]	Loss-on-ignition (%)	Exchangeable cations (m-equiv 100 g^{-1})				Acetic-acid extracted elements (µg g^{-1})			Exchangeable Ni (µg g^{-1})	CEC (m equiv 100 g^{-1})
			K	Na	Ca	Mg	P	Co	Cr		
280	5.7±0.1	12.4±0.8	0.14±0.01	0.10±0.01	7.7±1.1	24.6±1.4	4.1±0.4	2.5±0.3	0.5±0.1	13±2	49±3
610	6.0±0.2	23.9±3.3	0.42±0.08	0.17±0.03	12.4±2.8	10.6±1.0	7.1±1.3	4.1±0.6	0.5±0.2	15±2	102±3
870[2] (mor)	4.0±0.1	52.0±6.2	0.53±0.06	0.41±0.07	1.2±0.3	5.6±0.8	16.8±3.4	2.7±0.5	1.0±0.2	1.6±0.2	105±2
870[2] (mineral layer)	5.3±0.1	12.1±0.6	0.06±0.01	0.05±0.01	0.17±0.05	1.6±0.2	1.1±0.5	5.6±0.8	1.0±0.2	1.9±0.4	104±2

[1] in H_2O

[2] the combined mor and mineral layers' depth was 15 cm, but the depth of each layer varied between samples

280 m and 870 m, but the soil at 610 m was more calcareous. Exchangeable Ni showed relatively high concentrations at 280 m and 610 m.

5 Method

The cotton strip assay was first carried out in July–August 1983. However, the sampling periods of 24 and 34 days used then were too long, as many strips were decomposed well beyond the point when tensile strength (TS) measurements are valid. We repeated the work in 1984 with shorter sampling periods, and this timing proved to be more satisfactory. Only the 1984 work is reported in detail. In July 1984, 4 strips were placed at random within each of 10 subplots in each of the plots at 280 m, 610 m and 870 m. The methods followed those of Latter and Howson (1977). The strips were inserted vertically in the soil after removing the surface litter (ie recognizable plant remains but not soil organic matter). Sometimes (at 610 m and 870 m), the soils were too stony to allow the strips to be inserted to the full depth. The cotton strips were carefully removed from each subplot after 14 days and 20 days (610 m) or 21 days (280 m and 870 m). Two field control strips were used on each retrieval date at each plot. The strips were returned to the base camp, washed in tap water, air dried using a fan, and then stored between sheets of newspaper for 8 weeks. Some cloth control strips were kept at the Institute of Terrestrial Ecology's Merlewood Research Station to allow changes caused by transport, washing and storage to be assessed. After cutting, fraying and drying, the TS of 4 cm substrips frayed to 3 cm was tested at a relative humidity of 60–70%, using a Monsanto Type W tensometer with 5 cm wide pneumatic jaws.

The method of Hill et al. (1985) was used to calculate CT50, the time in days for the strips to lose 50% of their TS. This method allows comparisons between soils where the cotton has not been buried for the same length of time. First, the cotton rottenness function (CR) was calculated using the following formula:

$$CR = \sqrt[3]{\frac{\text{field control TS} - \text{final TS}}{\text{final TS}}}$$

where the field control TS is the mean TS of the 20 substrips from the 4 field control strips which had been placed (2 on each sampling date) in each site, and where the final TS is the median final TS of the samples for each depth, sample date and plot. (The median is used to avoid the bias due to non-linearity of change in TS with time and, as will be seen later, is further justified here in view of the high variability of the data.)

The CT50 was calculated using the formula:

$$CT50 = \frac{\text{no. of days in soil}}{CR}$$

6 Results

The cloth control strips had a mean TS of 46.7 ± 0.9 kg, and the field control strips a mean TS of 42.6 ± 0.8 kg. These values are significantly different (P<0.001) and demonstrate a TS loss during transport, storage and burial in the soil.

The experimental strips showed a very high within-plot variability (see Lindley & Howard 1988) between replicates, particularly at the lower altitudes. The results (Table 4) show that decomposition rates are ranked in the order 610 m, 280 m and 870 m. The data are simplified (by combining results for the 14 day and 20 or 21 day samples) in Figure 3. The plots at 280 m and 610 m show fairly similar patterns, whilst at 870 m decomposition is much slower and decreases markedly with depth.

The overall CT50 values are generally rapid compared with those in temperate areas (Hill et al. 1985; Ineson et al. 1988).

Table 4. The median tensile strength (TS) and CT50 (time in days to 50% loss of tensile strength, calculated by a method explained in the text) of cotton strips inserted to 20 cm depth (from the soil surface scraped free of litter) for 14, 20 or 21 days in 3 plots at different altitudes on Gunung Silam, Sabah. Field control value = 42.6 ± 0.8 kg

Altitude (m)	Sample dates (1984)	Days in field	Sample depth (cm)	No. of strips lost	No. of strips eaten by termites	No. of strips tested	Median TS (kg)	CT50
280	24 July–7 August	14	0–4	0	1	19	25.0	16.1
			4–8			19	25.0	16.1
			8–12			19	30.0	19.3
			12–16			19	31.0	20.1
			16–20			19	37.0	28.3
280	24 July–14 August	21	0–4	2	2	16	13.0	16.2
			4–8			16	20.5	20.8
			8–12			16	18.3	19.4
			12–16			16	20.3	20.7
			16–20			16	27.5	26.3
610	26 July–9 August	14	0–4	0	2	18	20.8	13.6
			4–8			18	21.3	13.8
			8–12			18	14.8	11.2
			12–16[1]			12	25.0	15.5
			16–20[1]			6	28.0	17.0
610	26 July–15 August	20	0–4	0	1	19	12.0	14.5
			4–8			19	–7.5	11.8
			8–12			19	14.5	15.9
			12–16[1]			17	23.8	21.2
			16–20[1]			8	17.8	17.6
870	25 July–8 August	14	0–4	0	1	19	38.0	27.9
			4–8			19	40.5	36.4
			8–12			19	42.0	52.4
			12–16			19	44.0[2]	—[2]
			16–20[1]			15	42.5	73.0
870	25 July–15 August	21	0–4	1	0	19	23.0	22.1
			4–8			19	26.0	24.3
			8–12			19	37.0	38.9
			12–16			19	42.0	78.6
			16–20[1]			14	36.0	36.6

[1] Several strips could not be buried to these depths

[2] Because the value for median TS did not differ from that of the field controls, the CT50 would be infinity. Only 21-day data were therefore used in Figure 3

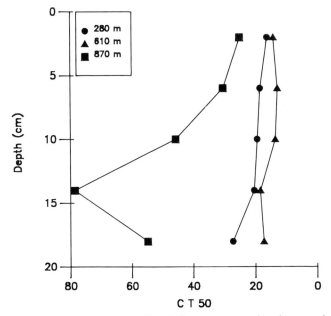

Figure 3. The mean CT50 for cotton strips inserted for 14 days and 21 days at 3 plots on Gunung Silam, Sabah. Supporting information is given in Table 4

7 Discussion

On Gunung Silam, a clear altitudinal trend in decomposition rates is lacking as the intermediate plot at 610 m had the highest rates. The differences are greatest between the plots at 610 m and 870 m, and yet the latitudinal difference is least between these 2 plots. Our 1983 data showed a similar pattern, in that the cotton strips lost TS more slowly at 870 m but there was no clear difference in that year between the results for 280 m and 610 m. The causes of the different rates (Table 4 & Figure 3) are probably complex. Some climatic factors (associated with the frequent cloud cap), such as solar radiation, change abruptly above 610 m. The rainfall patterns are very different between the summit and the lower parts of the mountain, and the former receives less rainfall (at least during much of July–September 1983 and 1984). The tree family composition at 870 m (Table 1) contrasts with those of the other 2 plots, and there is likely to be an important difference in litter quality which results in the observed slower rates of decomposition of the cotton strips and the accumulation of mor humus there. Leakey and Proctor (1987) have shown that the soil invertebrate biomass is much lower at 870 m (1.5 g m^{-2} alcohol wet weight) than at 610 m (4.5 g m^{-2}) or 280 m (7.1 g m^{-2}). Earthworms accounted for only 0.3% of the biomass at 870 m, 18% at 610 m and 85% at 280 m. These differences must reflect substrate quality, but are difficult to reconcile with the high rate of decomposition at 610 m. The lower plots are less acid and more calcareous than that at 870 m, and the distinctly calcareous plot at 610 m may favour a more active soil microflora. However, there is a higher mean loss-on-ignition (Table 3) at 610 m than 280 m. The results support the hypothesis that relatively low decomposition rates (and hence release of nutrients by mineralization) are likely to occur at higher altitudes on tropical mountains. The importance of this slower process on forest stature and composition needs further investigation, and it is apparently not the explanation of the substantial differences in forest stature between the plots at 280 m and 610 m. The cotton strip assay is seen to be a useful preliminary tool for investigating forest changes on mountains, and has led to the suggestion that, even along a forest continuum, the causes of the changes may differ from one part to another. The assay serves as a guide for the more detailed work which is a prerequisite for a full explanation of rainforest altitudinal zonation.

8 Acknowledgement

We are grateful to the Sabah Government for permission to work in their State, and to the Sabah Forest Department for their help and encouragement throughout. Datuk H J K M Mastan and Mr T C Liew, in particular, are thanked for their support. We thank ITE Merlewood for the use of the TS testing facilities.

9 References

Fox, J.E.D. 1978. The natural vegetation of Sabah, Malaysia. 1. The physical environment and classification. *Trop. Ecol.*, **19**, 218-239.

Grubb, P.J. 1971. Interpretation of the 'Massenerhebung' effect on tropical mountains. *Nature, Lond.*, **229**, 44-45.

Grubb, P.J. 1977. Control of forest growth and distribution on wet tropical mountains with special reference to mineral nutrition. *Ann. Rev. Ecol. Syst.*, **8**, 83-107.

Hill, M.O., Latter, P.M. & Bancroft, G. 1985. A standard curve for inter-site comparison of cellulose degradation using the cotton strip method. *Can. J. Soil Sci.*, **65**, 609-619.

Ineson, P., Bacon, P.J. & Lindley, D.K. 1988. Decomposition of cotton strips in soil: analysis of the world data set. In: *Cotton strip assay: index of decomposition in soils*, edited by A.F. Harrison, P.M. Latter & D.W.H. Walton, 155-165. (ITE symposium no. 24.) Grange-over-Sands: Institute of Terrestrial Ecology.

Latter, P.M. & Howson, G. 1977. The use of cotton strips to indicate cellulose decomposition in the field. *Pedobiologia*, **17**, 145-155.

Leakey, R.J.G. & Proctor, J. 1987. Invertebrates in the litter and soil at a range of altitudes on Gunung Silam, a small ultrabasic mountain in Sabah. *J. trop. Ecol.*, **3**, 119-129.

Lindley, D.K. & Howard, D.M. 1988. Some statistical problems in analysing cotton strip assay data. In: *Cotton strip assay: an index of decomposition in soils*, edited by A.F. Harrison, P.M. Latter & D.W.H. Walton, 25-27. (ITE symposium no. 24.) Grange-over-Sands: Institute of Terrestrial Ecology.

Proctor, J. & Woodell, S.R.J. 1975. The ecology of serpentine soils. *Adv. ecol. Res.*, **9**, 255-366.

Proctor, J., Lee, Y.F., Langley, A.M., Munro, W.R.C. & Nelson, T. 1988a. Ecological studies on Gunung Silam, a small ultrabasic mountain in Sabah. I. Environment, forest structure and floristics. *J. Ecol.* In press.

Proctor, J., Phillipps, C., Duff, G.K., Heaney, A. & Robertson, F.M. 1988b. Ecological studies on Gunung Silam, a small ultrabasic mountain in Sabah. II. Some forest processes. *J. Ecol.* In press.

Decomposition rates in the ericaceous belt of Mount Aberdare, Kenya

H COLLINS[1], H GITAY[2], E SCANDRETT[3] and T PEARCE[4]

[1]Department of Soil Science, University of Aberdeen, Aberdeen
[2]Department of Microbiology & Botany, University of Aberystwyth, Aberystwyth
[3]Hill Farming Research Organisation, Penicuik
[4]Department of Plant Science, University of Aberdeen, Aberdeen

1 Summary

Cotton strip decomposition rates were measured for 7 sites, at 3000–4000 m altitude, in the ericaceous belt of Mount Aberdare, Kenya.

Sites at lower altitudes (3100 m), estimated to have been undisturbed for the longest periods of time, showed a significantly greater rate of cellulose decomposition in the upper 8 cm of the soil than at a depth of 8–16 cm. Other sites did not show this trend.

There was no significant correlation between cellulose decomposition rates and soil concentrations of nitrogen (N), phosphorus (P), potassium (K), iron (Fe) or magnesium (Mg), although the relationship between decomposition rates and soil nitrogen in the upper 8 cm of the soil merited further investigation. Neither was there any significant correlation between cellulose decomposition rates and soil organic matter, or soil pH.

2 Introduction

The mountains of East Africa form a series of 'islands' of high altitude (>3000 m) in an otherwise tropical plateau at 1000–2000 m. Because of their relative isolation, they form a distinct group, with characteristic vegetation and a high degree of endemism, 81% at the species level (Hedberg 1961). Climatic features are important in creating this characteristic vegetation, especially the low temperatures (ground frost every night in most areas during the period of investigation), large diurnal temperature variations (up to 42°C recorded), large variations in monthly rainfall (an average of 19 mm in the driest month, 213 mm in the wettest month) and a 12-hour day.

Mount Aberdare comprises a range of rolling hills at 3000–4000 m above sea level (asl), with 2 peaks at the north and south ends (Sattima 4100 m; Kinangop 4231 m). The zonation at this altitude is complex, with 2 zones recognized (Hedberg 1951); the moorland shrub zone, consisting of tussock grassland with scattered shrubs and trees, and the ericaceous zone, dominated by arborescent and shrubby bushes of 4 genera which have an ericoid habit (*Erica*, *Philippia*, *Stoebe* and *Cliffortia*.)

The purpose of the study was to determine patterns in the vegetation of the ericaceous zone, and to attempt to assess the effect of various factors (both climatic and edaphic) on the zonation of the area. One of the soil parameters investigated was the rate of cellulose decomposition by the cotton strip assay (Latter & Howson 1977).

3 Site descriptions

Site 1: a gentle, south-facing slope, altitude 3100 m asl, with a stand of degenerate *Cliffortia* woodland (canopy closed, height 4 m), and a full ground cover of forbs and grasses. Estimated age of community — 50 years.
Soil — clay, dark reddish brown, deep, well-developed crumb above, massive below. Mottles of sesquioxides in upper horizons. pH 5.4.
Soil group, humic andosol.

Site 2A: the steep S/SW-facing slope of a stream gully, 3100 m asl, 16 m above stream bed, with a stand of mature *Philippia* woodland (canopy closed, height 8 m). Ground cover of mosses, grasses and forbs. Estimated age of community — 50 years.
Soil — clay, reddish brown, moderately deep, well-developed crumb above, massive below. Ferric horizon at 19–25 cm depth. pH 5.8
Soil group, humic andosol.

Site 2B: the S/SW-facing steep slope of a stream gully, 3100 m asl, 14 m below site 2A, with a stand of mature degenerate *Philippia* woodland (canopy closed, height 8 m) and a full ground cover of forbs and grasses. This site showed signs of regular disturbance by buffalo and other animals. Estimated age of community — 50 years.
Soil — silty clay loam, black to brown below, gleyed trachytic tuff below 100 cm, well-developed crumb above, weak crumb below. Mottles of sesquioxides. pH 5.2.
Soil group, humic andosol.

Site 3: a moderately steep, E/SE-facing exposed slope at 3500 m asl, with a stand of young *Cliffortia* and *Stoebe* (open canopy, bush height 0.5 m) and a ground cover of tussocky grasses. Evidence of recent burning. Estimated age of community — 2 years.
Soil — humic clay, brownish black, strongly developed crumb, strongly weathered trachyte below 90 cm depth. pH 5.7.
Soil type, dystric histosol.

Site 4A: the upper slope of a moderately steep west-facing valley, 3100 m asl, with mature *Philippia* woodland (canopy closed, height 14 m) and a full ground cover of forbs and grasses. Estimated age of community — 200 years.
Soil — clay loam, dark reddish brown, deep, well-developed crumb, streaks of iron in upper horizons. pH 5.3.
Soil group, humic andosol.

Site 4C: the crown of a slope, 3100 m asl, 10 m above site 4A with mature/degenerate *Stoebe/Phillipia* woodland (canopy open, height 1 m) and incursion of St John's wort (*Hypericum*), bamboo (*Bambusa*) and *Clutea*. Estimated age of community — 200 years.
Soil — silt loam, brownish black to reddish, brown below, deep, well-developed crumb, streaks of ferric material in upper horizons. pH 5.2.
Soil group, humic andosol.

Site 5: an east-facing moderate slope at the base of a basalt scarp, 4000 m asl, with a stand of young *Stoebe* (canopy closed, height 1 m) and a ground cover of grasses and few herbs. Estimated age of community — 4 years.
Soil — peat above to peat loam below, reddish black, strong crumb, trachyte bedrock below 100 cm depth. pH 5.4.
Soil group, dystric histosol.

4 Method

The period of study was January–March 1985, during one of the 2 annual dry seasons. Five main sites were chosen for investigation, and decomposition rates were determined for a total of 7 subsites.

The sites each measured 750 cm², and cotton strips (4 cm x 20 cm) were inserted randomly over the site to a depth of 16 cm below the litter layer. Strips were removed after 6 weeks, air dried and returned to Britain. The top section of the strips, which had been in the litter layer, were removed. The strips were then cut in half to correspond to 0–8 cm soil depth and 8–16 cm soil depth, and tensile strength was measured at the Institute of Terrestrial Ecology's Merlewood Research Station, using a tensometer (Latter & Howson 1977). Field controls were performed by inserting strips and removing them immediately at the time of removal of test strips. These then underwent all treatments performed on the test strips.

Total concentrations of selected ions in soils (N, P, K, Mg, Fe) were determined by sulphuric/perchloric acid digestion (Cresser & Parsons 1979), and results were confirmed for N by performing Kjeldahl digestions on random soil samples (Hanna 1964). Organic matter content was determined by weight loss-on-ignition at 375°C (Ball 1964). pH was determined by diluting fresh surface soil samples 1:3 with distilled water, and inserting a digital pH meter for readings after 15 minutes.

Soils were classified according to the FAO/UNESCO system, as used by the Kenya Soil Survey (1982).

5 Results and discussion

Tensile strength loss (CTSL) of cotton strips is assumed to be caused by microbial decomposition of cellulose fibres, and therefore is taken as a general indicator of the potential for cellulose decomposition in the soil.

The CTSL values (Figure 1) were examined by one-way analysis of variance (ANOVA) of pooled results, using site and depth as constants, but no significant difference was found between CTSL results at different depths or at different sites. However, based on tree girth and species composition, 4 sites (1A, 2A, 4A, 4C) were considered to have been undisturbed for a long period of time. On these sites, the CTSL at the 0–8 cm depth was significantly higher than at the 8–16 cm depth (P<0.01). By contrast, site 3, at 3500 m asl, recently burnt and very exposed, showed significantly lower cellulose decomposition in the surface soil than at the lower depth (P<0.01).

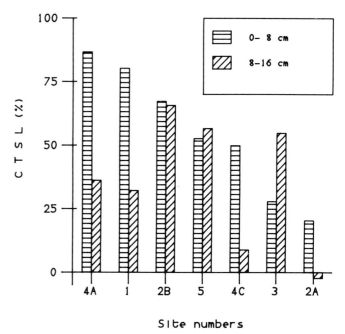

Figure 1. Tensile strength loss (CTSL) at the 7 sites (n = 5)

Sites 2B (2 m above a stream, and regularly disturbed by buffalo) and 5 (at 4000 m asl, but relatively unexposed to climatic extremes) both showed no significant difference in decomposition rates between the 2 soil depths.

The 3 sites with the greatest decomposition rates in the top 8 cm of soil (1, 2B and 4A) were all closed-canopy woodland sites which experienced the least diurnal temperature fluctuation of all sites (temperatures never falling to below 2°C or rising above 25°C), and all had a rich ground flora of forbs. This finding concurs with the observation that enzymic activities

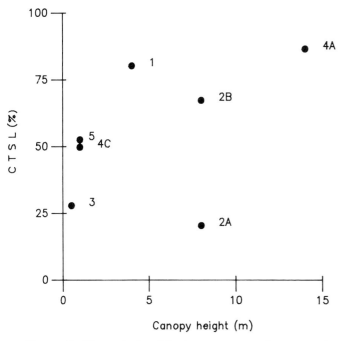

Figure 2. The relationship between tensile strength loss in the top 8 cm and canopy height at the 7 sites

levels in the lower 8 cm depth (Figure 3) or between cellulose decomposition rate and total soil P, K, Fe, organic matter content, or soil pH.

Because of the wide annual climatic variation in the area of study, it was not possible to predict annual cellulose decomposition rates from the present limited data. Results showed that the rate of cellulose decomposition did vary both within and between sites, but the reasons for this variation are unclear. It must be noted that cellulose decomposition rates, as determined by cotton strip assay, do not necessarily reflect overall microbial decomposition rates in soils.

in soils may be negatively related to the temperature range of sites, ie the highest activities occur where there is least extreme variation in temperature (Speir 1977).

There appeared to be an association between CTSL in the surface region and canopy height (Figure 2, r = 0.613), with the exception of site 2A which is in a stream gully. There also appeared to be a trend of increasing cellulose decomposition rates with increasing soil N levels in the upper 8 cm of the soil (Figure 3, r = 0.620); however, this relationship was not statistically significant if data from sites 4A and 1 were included. It is suggested that this aspect would benefit by further investigation. There was no significant correlation between decomposition rates and soil nitrogen

6 Acknowledgements

This work was carried out as part of the Aberdeen/Aberystwyth Expedition to the Aberdares, Kenya. We acknowledge all sponsors for their financial support, and the Institute of Terrestrial Ecology's Merlewood Research Station for assistance in the present study.

7 References

Ball, D.F. 1964. Loss-on-ignition as an estimate of organic matter and organic carbon in non-calcareous soils. *J. Soil Sci.*, **15**, 84-92.
Cresser, M.S. & Parsons, J.W. 1979. Sulphuric-perchloric acid digestion of plant material for the determination of N, P, K, Ca and Mg. *Analytica chim. Acta*, **109**, 431-436.
Hanna, W.J. 1964. Methods of chemical analysis in soils. In: *Chemistry of the soil*, edited by F.E. Bear, 474-502. 2nd ed. New York: Reinhold.
Hedberg, O. 1951. Vegetation belts of the East African mountains. *Svensk bot. Tidskr.*, **45**, 140-202.
Hedberg, O. 1961. The phytogeographical position of the Afroalpine flora. *Recent Adv. Bot.*, **1**, 914-919.
Kenya Soil Survey. 1982. *Exploratory soil survey report no. E1.* Nairobi: Kenya Soil Survey.
Latter, P.M. & Howson, G. 1977. The use of cotton strips to indicate cellulose decomposition in the field. *Pedobiologia*, **17**, 145-155.
Speir, T.W. 1977. Studies on a climosequence of soils in tussock grasslands. II. Urease, phosphatase, and sulphatase activities of topsoils and their relationships with other properties including plant available sulphur. *N.Z. Jl Sci.*, **20**, 159-166.

0-8 cm depth

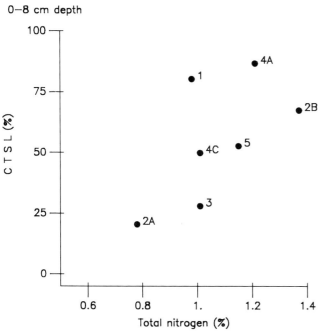

8-16 cm depth

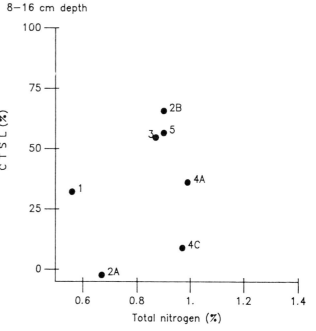

Figure 3. The relationship between tensile strength loss and total nitrogen at the 7 sites

Polar

Cotton strip decomposition in relation to environmental factors in the maritime Antarctic

D D WYNN-WILLIAMS
British Antarctic Survey, Cambridge

1 Summary
Potential cellulose decomposition in moss peat, measured as tensile strength loss (CTSL) of cotton strips inserted in the peat profile, was examined along a latitudinal transect in the maritime Antarctic. CTSL during a whole year at Signy Island (60°S), Galindez Island (65°S) and Rothera Point (68°S) was assessed relative to edaphic and microclimatic conditions. Occasional increases in tensile strength emphasized the importance of field controls. The pattern of CTSL down the profile varied between moss turves and carpets, and between sites. The increase in CTSL with depth in *Polytrichum* peat at Signy Island contrasted with a similar turf at Galindez Island and other tundra sites. Of the edaphic variables, nitrogen (N) and calcium (Ca) availability were probably contributory to CTSL limitation, along with extremes of pH, peat water content and associated anaerobiosis in stagnant conditions. However, the factor which appeared to explain CTSL differences between sites along the transect was sunshine and its associated ground-heating effect. This was the only variable evidently accountable for the high CTSL in the surface *Drepanocladus* peat at Rothera Point.

2 Introduction
The rate of moss peat decomposition at Signy Island, South Orkney Islands, Antarctica, is slow (Davis 1981), and results in the accumulation of peat in deep banks in drier localities (Fenton 1980) and in shallower carpets in wetter areas (Collins 1976). Measurement of the weight loss of autochthonous moss tissue in litter bags inserted in Signy Island moss peat is too slow (2–3% in 2 yr) to provide a sufficiently sensitive assay to detect seasonal and inter-species differences in decomposition rate (Davis 1986). However, the moss tissue contains 39–47% (dry wt) holocellulose, decreasing with depth in the peat profile (Davis 1986). Measurement of the decomposition of moss cellulose, therefore, provides a method for investigating a major aspect of peat accumulation. However, the natural cellulose of plant cell walls differs both in physical structure and chemical integrity with other cellular components from the crystalline nature of processed textiles such as cotton (Sagar 1988; Howard 1988). Nevertheless, the method of Latter and Howson (1977) for determining tensile strength loss of cotton strips inserted in the peat profile is appropriate for comparing the influences of different factors on the decomposition of the natural analogue under field conditions.

Despite the artificial nature of processed cotton cellulose, its field degradation at Signy Island (Heal *et al.* 1974; Wynn-Williams 1980) parallels measurements of moss tissue decomposition in litter bags (Baker 1972; Collins 1976; Fenton 1978; Davis 1986). Therefore, the cotton strip assay was chosen for the present investigation to compare the influence of moisture, temperature and other microclimatic and edaphic factors on potential cellulose decomposition at a range of sites in the maritime Antarctic in the summers of 1977–78 and 1980–81. This study paralleled investigations of peat respiration (Wynn-Williams 1984) and the microbiology of peat and mineral fellfield soils (Wynn-Williams 1985a).

3 Study sites
Two communities were studied at Signy Island, one a moss turf (SIRS 1) dominated by *Polytrichum alpestre* and *Chorisodontium aciphyllum,* and the other a moss carpet (SIRS 2) dominated by *Drepanocladus uncinatus, Calliergon sarmentosum, Calliergidium austro-stramineum* and *Cephaloziella varians* (Collins *et al.* 1975). *Polytrichum* and *Drepanocladus* communities occurred in the maritime Antarctic on a transect from 54°S to 68°S.

SIRS 1 and SIRS 2 have been described by Tilbrook (1973) and Collins *et al.* (1975). The physical and edaphic characteristics of these Signy Island sites and those for comparison near Faraday Station on Galindez Island and at Rothera Point (Figure 1) are given in Table 1. The *Polytrichum* moss turf on Galindez Island was on a gentle south-west slope at the base of a scree on Woozle Hill. The *Drepanocladus* carpet was in a wet runnel between these *Polytrichum* turves, close to the shoreline. The depth of peat underlying the *Polytrichum* turf and *Drepanocladus* carpet was 300 mm and 150 mm respectively.

At Rothera Point, where *Polytrichum* was absent, the *Drepanocladus* carpet was located in the west-facing catchment area of a small stream, at the foot of a snowfield 75 m from the British Antarctic Survey (BAS) research station. The peat depth was 30–40 mm.

4 Method
The procedures for insertion, retrieval and treatment of cotton strips are described in Latter and Howson (1977) and Wynn-Williams (1980). At least 10 replicate cotton strips were inserted vertically into the peat profile to a maximum depth of 180 mm. However,

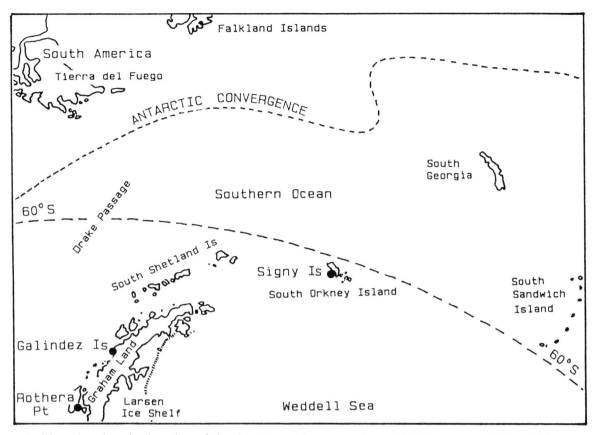

Figure 1. Map showing the location of the 3 cotton strip assay sites at BAS research stations (●)

Table 1. Habitat characteristics of the cotton strip insertion sites. Gravimetric data are expressed on a dry weight basis

Characteristic	Moss carpet communities			Moss turf communities	
	Signy Is	Galindez Is[1]	Rothera Pt	Signy Is	Galindez Is
Latitude	60°43'S	65°15'S	67°43'S	60°43'S	65°15'S
Duration of 1981–82 summer (d)	125	90	106	125	90
Total annual precipitation (mm)	282[2]	430[3]	c.710[3]	282[2]	430[3]
Mean ± SE (n) annual insolation (h)	561±78(33)	840±122(29)	1150±141(4)	561±78(33)	840±122(29)
Thawed period of strip insertion					
Duration (d)	127[2]	102[3]	123[3]	127[2]	102[3]
Mean air temperature (°C)	+0.58	−0.21	+0.05	+0.68	−0.21
Mean daily insolation (h)	2.01	2.73	3.11	2.01	2.73

Dominant moss stand	Calliergon			Drepanocladus	Drepanocladus	Polytrichum			Polytrichum
Depth (mm)	0–30[4]	40–60	Correlation coefficient[5]	30–80	30–80	0–30[4]	40–60	Correlation coefficient[5]	30–80
Peat field capacity (%)	1130	510	−0.84	900	900	490	520	−0.36	490
Peat moisture[6] (%)	886	588	−0.88	525	1162	442	496	−0.02	471
Percentage field capacity (%)	78	115	−0.42	58	129	90	95	0.24	96
pH	4.9	4.8	−0.10	4.7	5.1	4.3	4.3	0.13	4.1
Loss-on-ignition (%)	96	91	−0.55	89	87	97	97	−0.70	98
Total N (%)	1.81[7]	2.37[7]	0.86	3.07	3.31	0.72[7]	0.77[7]	0.67	0.55
NO_2^- + NO_3-N (mg 100 g^{-1})	NA	0.48	−0.63	0.01	2.3	0.12	0.074	0.88*	0.1
NH_4-N (mg 100 g^{-1})	29	61	−0.22	34	53	38	52	0.35	10
PO_4-P (mg 100 g^{-1})	31	20	−0.87*	16	9.5	59	23	−0.87*	8.5
K (mg 100 g^{-1})	300	23	−0.71	26	120	320	85	−0.75	35
Ca (mg 100 g^{-1})	140	60	0.92**	50	220	66	90	0.83*	88

[1] Mineral nutrient and LOI data from adjacent Uruguay Is
[2] 1977–78
[3] 1981–82
[4] Including green shoots
[5] With depth (* P<0.05, ** P<0.01) for the whole growing season
[6] At time of insertion of strips (February–March), depth 1–3 cm
[7] Davis (1986)
NA, not available

many were pulled out by skuas (*Catharacta skua*) during the summer. Determination of CTSL of sub-strips from defined depths down to 120 mm was made relative to both cloth controls and separate field controls for each community. The latter were removed immediately after insertion and treated in the same manner as test strips. At Signy Island in 1977–78, substrips were 20 mm wide from 10–30 mm below the moss surface (ie immediately below the green shoot layer) and 25 mm wide from each of the 6 lower depths. At the other sites, all substrips were cut 20 mm wide throughout the profile. All measurements of tensile strength (TS) were made at 100% relative humidity (RH), except for those strips from Signy Island in 1977–78, tested at 80% RH. TS data were adjusted for 100% RH and were converted to percentages of field control TS. This procedure enabled the comparison of data (n = 11) for unbleached calico (TS ± SE = 19.00 ± 0.10 kg cm^{-1}), used at Signy Island in 1977–78, with the Shirley Soil Burial Test Fabric (Walton & Allsopp 1977) used in 1980–81 (TS ± SE = 24.08 ± 0.26 kg cm^{-1}, n = 19).

The TS of field controls differed significantly ($P<0.01$) from cloth controls, but there was no significant difference within pooled field control strips from *Polytrichum*, *Chorisodontium*, *Calliergon* and *Cephaloziella* stands, so an overall mean ± SE of 17.52 ± 0.35 kg cm^{-1} (n = 23) was used previously (Wynn-Williams 1980, 1985a). However, to standardize the results throughout the transect, field data were calculated relative to mean TS values for each equivalent horizon field control. Of these field controls, the only significant differences were between *Chorisodontium* and both *Polytrichum* ($P<0.05$) and *Cephaloziella* ($P<0.01$), and between *Calliergon* and *Cephaloziella* ($P<0.05$). The use of separate field controls gave more sensitive measures of significance than those obtained using the overall mean field control value (Wynn-Williams 1980, 1985a), but the trends were similar.

Removal of strips by skuas from the *Chorisodontium* community in summer necessitated the extrapolation of spring data (74 d only) and adjustment for the consequent overestimation of CTSL rate by an empirical factor of 1.9 derived from a sole remaining strip.

5 Meteorological and edaphic data

Data for mean air temperature and insolation were obtained from the BAS Meteorological Section (D W Limbert pers. comm.). Details of microclimate monitoring at Signy Island are described in Walton (1982). The duration of thawed conditions was monitored at Signy Island and interpolated from meteorological data at Galindez Island and Rothera Point.

Peat moisture content and pH (1:1 dilution) were determined at the time of strip insertion (February 1977 at Signy Island and March 1981 at Galindez Island and Rothera Point), and were typical of summer

conditions (Wynn-Williams 1980, 1984). Redox conditions were assessed using silver-plated brass strips inserted vertically in the peat profile (Wynn-Williams 1980) and by platinum wire electrodes at SIRS 2, Signy Island (Yarrington & Wynn-Williams 1985). Chemical analyses of soil were carried out at the Institute of Terrestrial Ecology's Merlewood Research Station, using standard procedures (Allen *et al.* 1974).

6 Results
6.1 Habitat conditions (Table 1)
The duration of the 1981–82 summer thaw shows that the relative length of growing season at each site is not dependent on latitude alone. As the thaw generally proceeds downwards in the profile, the period of thaw is somewhat shorter at lower depths, so that CTSL of strips may be underestimated there. The mean air temperatures are based on monthly data from November to February at each site. There was a large difference in total annual insolation between the sites. However, for much of the year, there was extensive snow cover and the heating of the moss surface is negligible. During summer, after the ground had thawed, snow cover was sporadic and thin, so that incident radiation was more likely to reach the surface. The mean daily insolation during this thawed period was, therefore, a better indication of ground heating likely to influence cellulose decomposition near the surface. Integrated degree-day temperature data for peat profiles were only available for Signy Island (Walton 1977, 1982).

In spring at Signy Island, there was a significant increase in percentage moisture in *Polytrichum* stands with increasing depth of peat (r = 0.95, $P<0.005$ in 1975, and r = 0.99, $P<0.001$ in 1976). However, this trend gradually reversed during summer to become statistically insignificant. In *Calliergon* and *Drepanocladus* stands at Signy Island, there was an apparent decrease in percentage moisture with increasing depth, which was due to the consistently high water content and increasing bulk density with depth. The trend was most pronounced in spring (r = −0.99, $P<0.02$). Peat pH changed seasonally in both communities at Signy Island, although SIRS 2 was always less acid than SIRS 1 (Wynn-Williams 1985b, c).

Total N was consistently more abundant in moss carpets than turves, although extractable NO_2- + NO_3-N and NH_4-N was variable between sites and stands (Table 1). The overall range of extractable phosphate (PO_4-P) in moss peat from this sub-antarctic and antarctic region was extreme, being from 2.3 mg 100 g^{-1} in *Calliergon* peat at South Georgia to 1500 mg 100 g^{-1} in *Drepanocladus* peat influenced by a large penguin (*Pygoscelis papua*) rookery on the Dion Islands (68°S) (R I Lewis Smith unpubl.). Amounts of NO_2- + NO_3-N at Signy Island increased with depth in *Polytrichum* and decreased gradually in *Drepanocladus* peat, while PO_4-P decreased in both stands. Ca increased with depth in all communities.

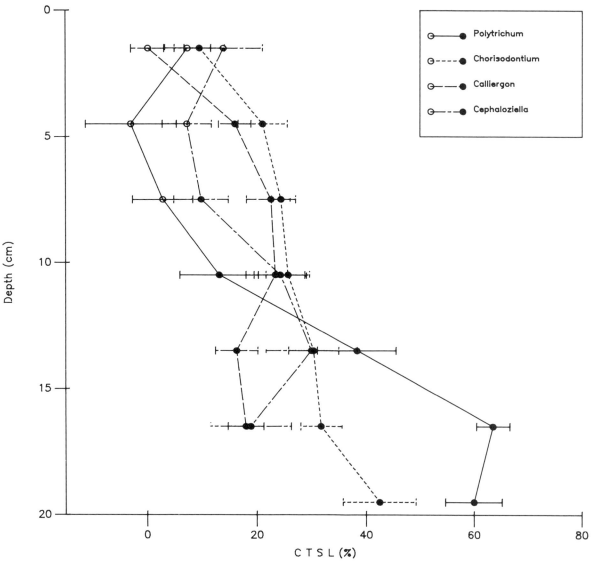

Figure 2. Tensile strength loss of cotton strips (CTSL) inserted in the peat profile of 4 moss communities at Signy Island. CTSL (%) values are corrected for a 100-d growing season. Solid circles indicate a significant (P<0.05 or better) difference from field controls. Field control values: Polytrichum 17.04 ± 0.63; Chorisodontium 19.08 ± 0.26; Calliergon 18.04 ± 0.23; Cephaloziella 15.51 ± 0.58. Chorisodontium data are derived from adjusted spring values (see Methods)

6.2 Tensile strength loss of cotton strips

For a given site, potential cellulose decomposition, as reflected by CTSL, could be compared using 3 parameters: (i) CTSL in the 10–30 mm horizon adjacent to the green shoots, (ii) the pattern of CTSL with increasing depth, and (iii) the depth of the peak in CTSL, if conspicuous.

At Signy Island (Figure 2), the CTSL in the 10–30 mm horizon was similar in all stands, irrespective of their habitats. However, the profile pattern of CTSL differs. In the 2 moss turf stands (Polytrichum and Chorisodontium), CTSL increased significantly with depth (r = 0.869, P<0.02, and r = 0.957, P<0.001, respectively). There were no such correlations in the moss carpet communities. A conspicuous peak in CTSL occurred near the bottom of the peat profile of moss turf, but there was no conspicuous maximum in moss carpet profiles.

6.3 Influence of temperature

The combined effect of temperature and duration of thaw on CTSL at different depths was investigated using the sum of degree-hours (Σdeg-h), ie the number of hours in which the temperature exceeded 0°C, multiplied by the temperature itself. Means were obtained for annual Σdeg-h values based on 10-day blocks of data from peat profiles at SIRS 1 and SIRS 2 in the summers of 1972–73 and 1973–74 (Walton 1977). The data were converted to degree-days and these means and the duration of thaw were compared with the time required for 50% CTSL of cotton strips during the 1977–78 summer at the same sites (Table 2). The negative trend between CTSL and thermal input was thermodynamically untenable, indicating that temperature was not the over-riding regulator of CTSL in the profile.

Table 2. Number of degree-days above 0°C required for 50% loss of tensile strength of cotton strips inserted in the moss peat profile at SIRS 1 and SIRS 2. CTSL values for 1977–78 were combined with mean annual Σdeg-d and duration of thaw data for 1972–73 and 1973–74

| Moss community | Depth (mm) | | | | | | Mean 10–180 | Correlation coefficient[2] |
	10–30	35–60	65–90	95–120	125–150[1]	155–180[1]		
Polytrichum	2288	—	3375	645	171	82	1312	−0.773
Chorisodontium	1763	628	387	328	216	163	581	−0.817
Calliergon	—	741	758	404	479	375	491	−0.772
Cephaloziella	914	1646	1056	389	260	357	770	−0.764

[1] Extrapolations based on the significant regression slopes of $\log_{10}$ transformed Σdeg-d on depth (mm)
[2] For Σdeg-d to 50% CTSL against depth. None were statistically significant
—, negligible or negative CTSL

6.4 Influence of moisture

The regression of mean percentage field capacity (% FC) for the peat profile at SIRS 1 and SIRS 2 during 1975–76 against the percentage CTSL of the same sites in the 1977–78 summer revealed no significant correlation between CTSL and moisture in any community. Moss turf was generally at field capacity during the summer, but moss carpets were frequently saturated. Rate limitation by moisture was not apparent under the field conditions investigated.

6.5 Influence of inorganic nutrients (Table 1)

Correlations between percentage CTSL and inorganic nutrients were in some cases depth-linked, and therefore inconclusive. Nevertheless, CTSL increased with NO_2-N + NO_3-N availability in *Polytrichum* turf (P<0.001) but not in *Chorisodontium* or in the N-richer moss carpet. The negative correlation of CTSL with PO_4-P in *Chorisodontium* and *Calliergon* peat was unlikely to be ecologically meaningful. A trend of increasing CTSL with increasing Ca was significant in *Chorisodontium* alone (P<0.01). The negative correlation between CTSL and K in *Chorisodontium* and *Calliergon* (P<0.02) was probably a depth effect. No other trends were significant.

6.6 Variation between sites

The difference between CTSL in the moss carpet stands was extreme in the 10–30 mm horizon at the 3 sites, being negligible at Signy Island, considerable at Galindez Island and extensive at Rothera Point. On extending comparisons of CTSL with edaphic variables for other antarctic sites (Figure 3), variations between stands and sites become apparent. Although CTSL in *Polytrichum* peat was low in the 10–30 mm horizon and was high near the bottom of the profile at Signy Island, it was negligible in *Polytrichum* at Galindez Island.

There were also marked differences in profiles of CTSL with depth between the same stands at different sites. The large increase in CTSL with depth in *Polytrichum* peat at Signy Island contrasted with minimal CTSL throughout the profile in this type of peat at Galindez Island. Apart from the 10–30 mm horizon, CTSL was moderate throughout the *Calliergon* peat

profile at Signy Island but maximal in the mid-profile of comparable *Drepanocladus* peat at Galindez Island.

6.7 Variations between moss communities

The overall peat decomposer activity in different moss communities can be compared broadly by the mean percentage CTSL for the whole profile. For *Polytrichum* peat at Signy Island, the mean (± SE) for 6 depths to 180 mm was 26.0 ± 10.5%, whereas for 8 depths to 240 mm in similar peat at Galindez there was an apparent mean TS increase of 1.8 ± 1.0%. In moss carpets, the percentage CTSL values at Signy Island, Galindez Island and Rothera Point were 16.1 ± 3.4%, 31.1 ± 9.1% and 46.6% for 6, 8 and 1 depths, respectively. This ratio of 1:2:3 parallels the values for decomposer activity in the 10–30 mm horizon alone, and is the converse of what might have been expected geographically.

7 Discussion

The CTSL in the present study revealed patterns of potential cellulose breakdown which differed between communities and sites and which were correlated with certain edaphic and microclimatic factors. Trends were partly obscured by apparent increases in TS at Signy Island and Galindez Island, probably due to stretching and the cementing action of peat organic materials such as mucopolysaccharides (French 1984, 1988). TS increases in maritime antarctic peat have been reported previously (Heal *et al.* 1974). If the peat is relatively dry during the insertion of field controls, the cementing materials may impregnate the strips less efficiently than under subsequent wetter conditions. Cementation of controls may, therefore, be less than in test strips, leading to an underestimation of CTSL. The method might be improved by inserting extra field controls in wet conditions.

Although the tensile strength loss of cotton as a standard substrate cannot reflect all aspects of peat decomposition, Schmidt and Ruschmeyer (1958) showed a correlation of 0.73 (n = 22, P<0.001) between soil CO_2-release and cellulose (cotton) decomposition in a range of loams. However, they found a rapid loss in the first few days of burial not matched by increased carbon dioxide (CO_2) release. This sensi-

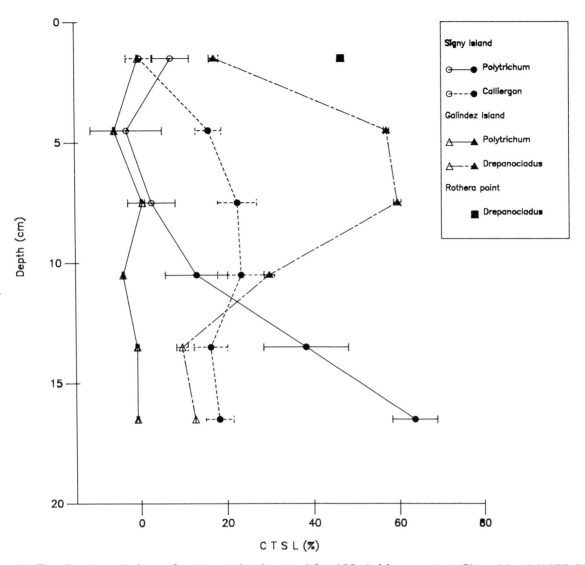

Figure 3. Tensile strength loss of cotton strips inserted for 100 d. Moss peat at Signy Island (1977–78), at Galindez Island and Rothera Point (1981–82). Solid circles indicate significant difference from field control (P<0.05 or less). Field control values: Signy Is, Polytrichum 17.04 ± 0.63, Calliergon 18.04 ± 0.23; Galindez Is, Polytrichum 22.63 ± 0.16, Drepanocladus 22.19 ± 0.23; Rothera Point, Drepanocladus 21.30 ± 0.65

tivity of the CTSL method is valuable for inter-site comparisons, and necessary at maritime antarctic sites where peat decomposition is slow (Davis 1986).

The low CTSL in the 10–30 mm horizon of *Polytrichum* peat at Signy Island was first reported at a similar site nearby (Hut Bank) in the 1970–71 growing season (Heal *et al.* 1974), and again in 1975–76 by Fenton (1978). This unusual feature, shared by *Polytrichum* peat at Galindez Island, was in contrast to most other tundra sites of the International Biological Programme (IBP) (Baker 1974; Heal *et al.* 1974, 1981) and a *Polytrichum* moss bank site at South Georgia (Smith & Walton 1988; Walton 1985). This contrast persists in the profile as CTSL tends to decrease with increasing depth in arctic tundra peat, such as palsa peat at Kevo (r = –0.934, P<0.02) (Baker 1974), rather than reach a maximum at 160–180 mm as occurred here. This depth-related increase was also observed at Signy Island in 1970–71 and 1975–76, but not in 1978–79 when CTSL was moderate near the surface (Davis 1986), although it did increase markedly from the

mid-profile downwards. Caution is, therefore, required when generalizing from individual seasons because of annual variations in climate.

Although temperature and moisture are fundamental factors for cellulose breakdown (Heal *et al.* 1974), their influence may be obscured by interactions with other rate-limiting variables. The conspicuous decrease in Σdeg-h down the peat profile at Signy Island was not paralleled by demonstrable changes in CTSL. The fact that CTSL near the surface was faster in *Chorisodontium* than *Polytrichum* may be associated with a higher mean surface temperature due to its hummocky growth form (Baker 1972), because other abiotic factors and nutrient levels were similar in both turf species. Although the summer estimate of the peat profile pattern for *Chorisodontium* extrapolated from spring data was valid, absolute values of percentage CTSL were tentative.

The moss turf moisture content rarely exceeded field capacity and was probably near optimal for tundra

(Heal *et al.* 1974), but the moss carpets were sufficiently wet (frequently >250% FC on a dry weight basis) to result in anaerobiosis if stagnant. Assuming that anaerobic breakdown of peat may be slower than aerobic decomposition (Davis 1981, 1986), these anaerobic conditions are reflected by the lower CTSL found at depth at SIRS 2 in *Calliergon,* and especially in waterlogged *Cephaloziella.* The *Cephaloziella* peat profile was sufficiently anaerobic for methanogenesis to occur (Yarrington & Wynn-Williams 1985).

Although both communities had a similar rate of net primary production, the moss carpet accumulated less peat than the turf, suggesting a higher overall decomposition rate in carpets (Davis 1981). This conclusion was not apparently supported by the present findings for CTSL at Signy Island. However, moss turf contains about 1.6 times more cellulose than moss carpet (Davis 1986), so less cellulose decomposition is required in carpets to prevent peat accumulation. The accumulation of peat in the moss turf relative to the moss carpet may be due to lower pH, lower availability of N, and a high C/N ratio (Davis 1986).

The lower decomposer activity of *Polytrichum* peat at Galindez Island relative to Signy Island may be due to N and P deficiencies, in view of their similar physical environments. At Signy Island, the increase in available NO_2-N + NO_3-N and Ca at greater depths may partly explain the decomposition gradient, irrespective of decreasing P and potassium (K) which were plentiful relative to other similar IBP tundra sites (Brown & Veum 1974). The correlation between CTSL and NO_3-N availability in N-limited *Polytrichum* turf reflects a similar relationship shown independently of depth in loam soils by Schmidt and Ruschmeyer (1958).

Comparison of CTSL in moss carpets at the 3 sites shows a wide diversity. CTSL in *Drepanocladus* peat was significantly greater than in *Calliergon* at all depths, especially at Rothera Point where the carpet was flushed with aerobic water which may have removed potentially toxic peat humic acids. Abiotic and nutritional factors were similar at all sites, except for large differences in sunshine during the thawed period. Sunshine during winter has little influence on ground heating because of its short duration and the high reflectivity of the snow cover. However, during the thawed period of summer, snow cover is sporadic, and Walton (1982) showed that up to 20% of photosynthetically active radiation can penetrate 20 cm of late-winter snow. Solar heating has already been proposed to explain part of the elevated decomposition rate in *Chorisodontium* elsewhere at Signy Island (Baker 1972). The heating effect of insolation on the 0–30 mm horizon of *Drepanocladus* peat at Rothera Point, although limited by snow cover and water content, may therefore have a significant influence on CTSL.

The rate of CTSL may be associated with size and density of microbial populations, as suggested for subantarctic sites by Smith and Walton (1988). Wynn-Williams (1980, 1985a) found a correlation between CTSL and the golden, potentially microbially induced, discoloration of cotton strips and concurrent isolation of potentially cellulolytic *Cytophaga* spp. He also reported a significant escalation of the total and gold chromogenic bacterial microflora at Rothera Point following the construction of the new BAS research station, which may have introduced or stimulated decomposer micro-organisms. These micro-organisms may be associated with the high CTSL found at this site.

8 Conclusions

With suitable controls and due consideration of the artificiality of the pure substrate, the cotton strip assay is sufficiently sensitive to compare the slow rates of microbial cellulolytic activity prevailing in soils of the maritime Antarctic. In this region, CTSL appears to be regulated partly by the chemical nature of the moss peat and its N, P, K and Ca content, but primarily by its moisture content and the heating effect of incident solar radiation.

The validity of extrapolating cotton strip assay data to predict decomposition rates for complex lignified plant tissue is debatable (Davis 1986), but the high proportion of cellulose in moss tissue at Signy Island makes the assay relevant to the estimation of peat decomposition. Calibration of the assay, using cotton strips buried after impregnation with indigenous moss peat homogenate and with standard broad-spectrum nutritional amendments such as lucerne meal (used for optimizing soil microbial activity for pesticide testing), may help to interpret cotton strip assay data.

9 Acknowledgements
I thank Dr J C Ellis-Evans, Dr P Griffiths, C Jeffers and M Evans for help with retrieval of the cotton strips. I am grateful to Drs W Block and D W H Walton for critical review of the manuscript. John Corkhill (BAS Meteorological Section) provided the meteorological data. Dr R I Lewis Smith provided the results of chemical analyses carried out at the Institute of Terrestrial Ecology's Merlewood Research Station.

10 References
Allen, S.E., Grimshaw, H.M., Parkinson, J.A. & Quarmby, C. 1974. *Chemical analysis of ecological materials.* Oxford: Blackwell Scientific.

Baker, J.H. 1972. The rate of production and decomposition of *Chorisodontium aciphyllum* (Hook. f. & Wils.) Broth. *Bull. Br. antarct. Surv.,* **27,** 123-129.

Baker, J.H. 1974. Comparison of the microbiology of four soils in Finnish Lapland. *Oikos,* **25,** 209-215.

Brown, J. & Veum, A.K. 1974. Soil properties of the international tundra biome sites. In: *Soil organisms and decomposition in tundra,* edited by A.J. Holding, O.W. Heal, S.F. MacLean & P.W. Flanagan, 27-48. Stockholm: Tundra Biome Steering Committee.

Collins, N.J. 1976. The development of moss peat banks in relation to changing climate and ice cover on Signy Island in the maritime Antarctic. *Bull. Br. antarct. Surv.,* **43,** 85-102.

Collins, N.J., Baker, J.H. & Tilbrook, P.J. 1975. Signy Island, maritime Antarctic. In: *Structure and function of tundra ecosystems*, edited by O.W. Heal & T. Rosswall, 345-374. (Ecol. Bull. no. 20.) Stockholm: Swedish Natural Research Council.

Davis, R.C. 1981. Structure and function of two Antarctic terrestrial moss communities. *Ecol. Monogr.*, **51,** 125-143.

Davis, R.C. 1986. Environmental factors influencing decomposition rates in two Antarctic moss communities. *Polar Biol.*, **5,** 95-103.

Fenton, J.H.C. 1978. *The growth of Antarctic moss peat banks.* PhD thesis, University of London.

Fenton, J.H.C. 1980. The rate of accumulation in Antarctic moss banks. *J. Ecol.*, **68,** 211-228.

French, D.D. 1984. The problem of 'cementation' when using cotton strips as a measure of cellulose decay in soils. *Int. Biodeterior.*, **20,** 169-172.

French, D.D. 1988. The problem of cementation. In: *Cotton strip assay: an index of decomposition in soils*, edited by A.F. Harrison, P.M. Latter & D.W.H. Walton, 32-33. (ITE symposium no. 24.) Grange-over-Sands: Institute of Terrestrial Ecology.

Heal, O.W., Howson, G., French, D.D. & Jeffers, J.N.R. 1974. Decomposition of cotton strips in tundra. In: *Soil organisms and decomposition in tundra*, edited by A.J. Holding, O.W. Heal, S.F. MacLean & P.W. Flanagan, 341-362. Stockholm: Tundra Biome Steering Committee.

Heal, O.W., Flanagan, P.W., French, D.D. & MacLean, S.F. 1981. Decomposition and accumulation of organic matter. In: *Tundra ecosystems: a comparative analysis*, edited by L.C. Bliss, J.B. Cragg, O.W. Heal & J.J. Moore, 587-633. Cambridge: Cambridge University Press.

Howard, P.J.A. 1988. A critical evaluation of the cotton strip assay. In: *Cotton strip assay: an index of decomposition in soils*, edited by A.F. Harrison, P.M. Latter & D.W.H. Walton, 34-42. (ITE symposium no. 24.) Grange-over-Sands: Institute of Terrestrial Ecology.

Latter, P.M. & Howson, G. 1977. The use of cotton strips to indicate cellulose decomposition in the field. *Pedobiologia*, **17,** 145-155.

Sagar, B.F. 1988. The Shirley Soil Burial Test Fabric and tensile testing as a measure of biological breakdown of textiles. In: *Cotton strip assay: an index of decomposition in soils*, edited by A.F. Harrison, P.M. Latter & D.W.H. Walton, 11-16. (ITE symposium no. 24.) Grange-over-Sands: Institute of Terrestrial Ecology.

Schmidt, E.L. & Ruschmeyer, O.R. 1958. Cellulose decomposition in soil burial beds. I. Soil properties in relation to cellulose degradation. *Appl. Microbiol.*, **6,** 108-114.

Smith, M.J. & Walton, D.W.H. 1988. Cellulose decomposition in four subantarctic soils. *Polar Biol.* In press.

Tilbrook, P.J. 1973. The Signy Island terrestrial reference sites. I. An introduction. *Bull. Br. antarct. Surv.*, **33 & 34,** 65-76.

Walton, D.W.H. 1977. *Radiation and soil temperatures 1972-74: Signy Island terrestrial reference site.* (British Antarctic Survey data no. 1.) Cambridge: British Antarctic Survey.

Walton, D.W.H. 1982. The Signy Island terrestrial reference sites: XV. Microclimate monitoring, 1972-74. *Bull. Br. antarct. Surv.*, **55,** 111-126.

Walton, D.W.H. 1985. Cellulose decomposition and its relationship to nutrient cycling at South Georgia. In: *Antarctic nutrient cycles and food webs*, edited by W.R. Siegfried, P.R. Condy & R.M. Laws, 192-199. Berlin: Springer.

Walton, D.W.H. & Allsopp, D. 1977. A new test cloth for soil burial trials and other studies on cellulose decomposition. *Int. Biodeterior. Bull.*, **13,** 112-115.

Wynn-Williams, D.D. 1980. Seasonal fluctuations in microbial activity in Antarctic moss peat. *Biol. J. Linn. Soc.*, **14,** 11-28.

Wynn-Williams, D.D. 1984. Comparative respirometry of peat decomposition on a latitudinal transect in the maritime Antarctic. *Polar Biol.*, **3,** 173-181.

Wynn-Williams, D.D. 1985a. Comparative microbiology of moss-peat decomposition on the Scotia Arc and Antarctic Peninsula. In: *Antarctic nutrient cycles and food webs*, edited by W.R. Siegfried, P.R. Condy & R.M. Laws, 204-210. Berlin: Springer.

Wynn-Williams, D.D. 1985b. The Signy Island terrestrial reference sites: XVI. Peat O_2-uptake in a moss turf relative to edaphic and microbial factors. *Bull. Br. antarct. Surv.*, **68,** 47-61.

Wynn-Williams, D.D. 1985c. The Signy Island terrestrial reference sites: XVII. Peat O_2-uptake in a moss carpet relative to edaphic and microbial factors. *Bull. Br. antarct. Surv.*, **68,** 61-70.

Yarrington, M.R. & Wynn-Williams, D.D. 1985. Methanogenesis and the anaerobic micro-biology of a wet moss community at Signy Island. In: *Antarctic nutrient cycles and food webs*, edited by W.R. Siegfried, P.R. Condy & R.M. Laws, 229-233. Berlin: Springer.

Using the cotton strip assay to assess organic matter decomposition patterns in the mires of South Georgia

G J LAWSON

Institute of Terrestrial Ecology, Merlewood Research Station, Grange-over-Sands

1 Summary

A large number of cotton strips were inserted on 4 mire types and a fellfield site on South Georgia during 1975 and 1976. Overlapping series of strips permitted an assessment of the effects of season, site, depth, duration of insertion and soil temperature on tensile strength loss. Whilst the sites demonstrate considerable differences, they are all amongst the most actively decomposing habitats in the tundra biome. Despite major seasonal changes, there is a strong indication of peak decomposition at depths of 8–14 cm. This pattern is unlike that reported for northern hemisphere mires, where decomposition peaks closer to the surface.

2 Introduction

Cotton strips have been used several times to assess potential organic matter decomposition patterns in different habitats on the South Atlantic island of South Georgia (54–58°S, 36–38°W). This paper presents research data on decomposition of cotton strips during 1974 and 1975 on 5 sites, typical of widespread mire and fellfield communities on the island.

3 Site description

The 5 selected sites are within or adjacent to the International Biological Programme (IBP) study area described by Lewis-Smith and Walton (1975). The climate of the sites includes sub-zero mean monthly temperatures for 4–5 months in the year, and around 170 snow-free days at sea level. The vegetation, altitude, water table and soil pH characteristics of these sites are summarized in Tables 1 and 2 (Plate 10). Further information, including maps and photographs of the 2 principal sites, is presented in Lewis-Smith (1981) and Lawson (1985).

Table 2. Soil temperature comparison (°C, spot thermistor measurements at noon, n = 27 at all sites)

| | | Depth (cm) | | | | | |
		2	5	10	20	40	Mean
Seepage	Mean	6.33	4.64	3.67	3.27	3.32	4.25
slope (SS)	Minimum	−0.85	−0.44	−0.06	0.27	0.61	−0.01
	Maximum	20.13	15.05	10.45	9.41	7.57	12.52
Basin bog	Mean	4.27	4.05	2.70	2.76	2.91	3.34
(BB)	Minimum	−3.18	−2.90	−2.18	0.09	0.33	−1.57
	Maximum	15.90	14.80	9.65	7.75	7.31	11.08
Steep	Mean	6.27	5.07	3.99	3.21	3.34	4.38
flush (SF)	Minimum	−1.39	−0.74	−0.21	0.18	0.24	−0.38
	Maximum	21.50	14.48	11.38	8.93	8.51	12.96
Gentle	Mean	4.54	4.60	3.69	3.12	3.18	3.83
flush (GF)	Minimum	−1.47	−1.47	−0.71	−0.09	0.33	−0.68
	Maximum	13.99	14.92	11.33	8.93	9.12	11.53
Fellfield	Mean	4.08	3.44	2.60	Rock	Rock	
(FM)	Minimum	−2.50	−2.18	−1.08			
	Maximum	13.04	11.33	8.06			

4 Materials and methods

The method of cotton strip preparation described in Latter and Howson (1977) was followed, using a domestic pressure cooker as an autoclave. The cloth used was the unbleached calico used for the earlier IBP studies (Heal *et al.* 1974).

A total of 540 strips were inserted in soil profiles, and seasonal differences were investigated using combinations of discrete and overlapping strip series (Figure 1). Greater numbers of strips were placed on the

Table 1. Summary characteristics of the 5 study sites

Site	Altitude (m)	Vegetation	Water table (cm)	pH	Notes
Seepage slope (SS)	85	Sparse *Rostkovia magellanica*	0–4	6.0	Deep snow, thin ice crust
Basin bog (BB)	60	Dense *R. magellanica* *Polytrichum alpestre* *Chorisodontium aciphyllum* *Cladonia* sp.	0–22	4.7	Shallow snow, thick ice crust, early melt
Steep flush (SF)	84	Dense *Juncus scheuchzerioides* *Calliergon sarmentosum* *Philonotis acicularis*	Rapid sub-surface flow	6·8	Deep snow, thin ice crust, late melt
Gentle flush (GF)	60	Sparse *J. scheuchzerioides* Sparse *Acaena magellanica* *Drepanocladus uncinatus*	Moderate water flow	5.9	Thin snow, thick ice crust, early melt
Fellfield (FM)	2	Stunted *R. magellanica* *Polytrichum alpinum* *Conostomum pentasticum* *Cladonia* sp.	Thin mineral soil, freely draining	5.7	Medium snow depth, late melt

Plate 10. Sites on South Georgia used for cotton strip assay:
i. seepage slope   ii. basin bog
(Photographs G J Lawson)

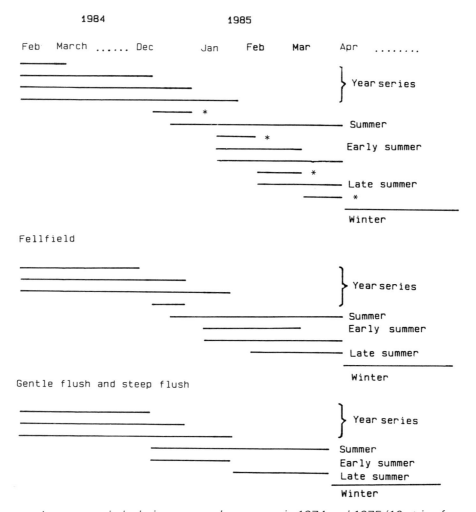

Figure 1. *Cotton strip assay periods during seasonal sequence in 1974 and 1975 (10 strips for each assay period)*

seepage slope and basin bog sites, where other studies of decomposition and nutrient cycling have been carried out (Lawson 1985).

Insertions were made at all sites using a flat-backed spade, leaving 6–8 cm of the strips exposed above the moss surface. This procedure could not be followed at the fellfield (FM) site, because of the shallow soil, and both ends of the strip were therefore inserted to increase replication and to anchor the cloth more firmly in position. Cloth control strips of washed and sterilized strips were retained, and each series included field control strips, which were inserted and removed immediately.

The cotton strips were carefully washed by hand, air dried, and cut into 4 cm wide substrips at depths related to the moss surface. On the 2 principal sites, it was also necessary to note the depth at which recognizable moss stems gave way to ombrogenous peat. This zone was presumably associated with high microbial activity. It occurred at a constant depth of 8 cm under the moss carpet of the seepage slope (SS), but below the irregular surface of *Polytrichum alpestre* turves at the basin bog (BB) its depth varied from 1 cm to 15 cm. This discontinuity was termed the limit

of recognizable vegetation (LRV), and was not present at the fellfield or flush sites.

Soil temperatures were not measured during the study period, but data for 1978 were available for all sites (Table 2).

5   Results
5.1   Between-site comparisons
5.1.1   *The complete year*
One series of cotton strips was inserted at all 5 sites in January 1974 and retrieved after approximately 12 months. Though decomposition proceeded well beyond the optimum for tensile strength measurements, differences in tensile strength loss of cotton (CTSL) could be shown between the sites (Figure 2 i). The basin bog (BB) demonstrates least overall CTSL, and a markedly lower loss at greater depths. The middle portions of strips from the other sites, particularly the steep flush (SF), were often completely disintegrated.

5.1.2   *The early summer*
This series of strips was inserted in mid-December 1974, after a late thaw. The assay period was 28 days at the basin bog, the seepage slope (SS) and the fellfield (FM) sites, but 65 days at the 2 flush sites (GF

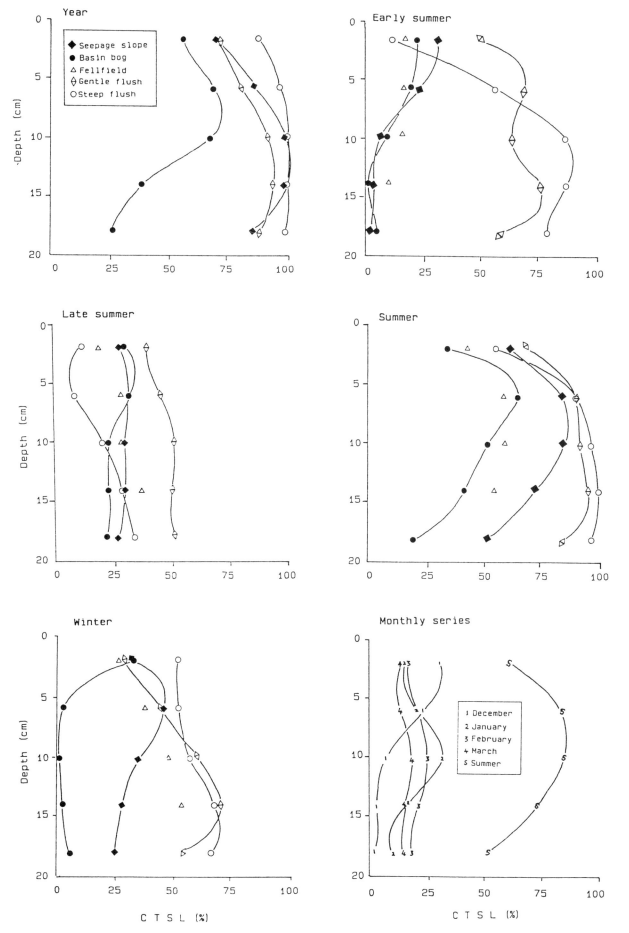

Figure 2. Patterns of tensile strength loss of cotton (CTSL) with depth in soil profiles over entire year and for the seasonal assay periods (Figure 1) at 5 sites on South Georgia. For the seepage slope site, the seasonal trends for 4 discrete assay periods (* in Figure 1) are also compared to an entire summer assay period

and SF). Data from the flush sites have, therefore, been adjusted linearly to express the data on the same 28 days' timescale as the other sites (Figure 2 ii). These 2 flush sites clearly show maximum decomposition at 10–14 cm during the first 2 months of summer. Over a shorter period after thaw, the other sites showed maximum decomposition close to the surface, and very little below 10 cm.

5.1.3 The late summer
The assay period for the late summer series was 54 days, starting on 21 February 1975. It showed much smaller differences in decomposition with depth in the profiles in all sites. There is an indication that CTSL is lowest close to the surface of the flush sites, and losses appear to increase slightly with depth in the fellfield, seepage slope and basin bog sites (Figure 2 iii). Water flows through both flushes at all depths, but the surface of the steep flush (SF) resembles a stream, which may explain the unusually low decomposition recorded for this site.

5.1.4 The complete summer
The assay period for all sites was 114 days, starting on 23 December 1974. The period covers approximately the sum of the periods displayed in Figures 2 ii and 2 iii, and might be expected to integrate rates of decomposition during the 2 periods, bearing in mind that the CTSL is not linear with time. There are some qualitative indications that the patterns of decomposition down soil profiles from this series (Figure 2 iv) do reflect the sum of Figures 2 ii and 2 iii.

Decomposition generally decreases with depth in early summer at the fellfield site, but increases in late summer. Over-summer strips record a maxima at 4–8 cm, which is consistent with combining the 2 trends.

The 60–70% CTSL near the surface of the steep flush does not, however, correspond to the smaller losses shown in Figures 2 ii and 2 iii. Similarly, the clear maxima at 6–10 cm in the seepage slope and basin bog sites do not conform well to the sum of early summer and late summer trends. Nevertheless, in both cases, the pattern of CTSL down the profile for the summer period matches that for the whole year (Figure 2 i).

5.1.5 Overwinter
Overwinter CTSL approaches 70% at 14–18 cm in the flush sites, despite low temperatures, but is insignificant below 6 cm in the basin bog. The seepage slope is intermediate between the other sites and shows a pattern of declining decomposition with depth (Figure 2 v).

5.2 Seasonal comparisons
Cotton strips were placed at the basin bog and the seepage slope sites to cover the entire summer in 4, approximately monthly, intervals. The data for the seepage slope are illustrated (Figure 2 vi), and show

that the sum of the monthly series does not necessarily match the pattern observed in the strips which remained in place over the whole summer. There is some evidence that the decomposition maxima are close to the surface immediately after thaw, but move down the profile as the season proceeds. At the end of the season, no significant differences with depth are apparent.

6 Discussion
This summary paper provides only a superficial review of the data available from one of the largest sets of cotton strips to be inserted at closely adjacent sites. These data are to be analysed after being reworked to a standardized timescale and temperature base in degree-days, and adjusted to account for the variable depth of turf-forming mosses. The details of these analyses will be reported elsewhere.

The general pattern of potential decomposition showed that the highest recorded CTSL was from 10–14 cm at the steep flush site, and exceeded 90% after only 65 days in the first half of summer. This rate is greater than the 100% CTSL in a 100-day period predicted for a South Georgian dwarf-shrub community by Smith and Walton (1988), and confirms their suggestion that peat formed from greater rush (Juncus scheuchzerioides) has a particularly active microflora (Smith and Walton 1985). The pH recorded from this site averaged 6.8, which may explain why the decomposition rate there is similar to that recorded from the surface of base-rich grassland in the UK and Ireland (64% loss in 52 days and 100% loss in 63 days, respectively (Heal et al. 1974)).

Heal et al. (1974) collated information from 24 IBP tundra biome sites, and it is interesting to note that only in the antarctic sites has there been any previous indication of CTSL increasing with depth. The present study shows that decomposition frequently peaks at depths between 8 cm and 14 cm. Even the fellfield site, with its comparatively dry mineral soil, shows the same trend of increasing decomposition with depth. Walton (1985) confirmed the above patterns in another South Georgian mire, but reported quite different trends for a dry grassland and moss bank. Whilst considerable differences exist between the 5 sites, even the comparatively cold and acid basin bog shows a higher rate of decomposition than 22 of the 24 IBP tundra sites (Heal et al. 1974).

Both the rate and pattern of decomposition undergo marked changes at different times of year. During thaw, for example, some sites retain greatest activity close to the surface (SS, FM and BB), whilst the flush sites (SF and GF) show a higher CTSL at greater depth. As the summer proceeds, the patterns of decomposition down soil profiles at all sites tend to become more similar.

Differences between the sites partially follow the observed patterns of temperature and pH (lowest at BB and highest at SF), but there are several anomalies which cannot be explained by temperature alone. For example, the GF and BB sites, because of shallower snow covers, have lower surface temperatures than the SF and SS sites, yet the 2 most active sites are SF and GF. Average temperature declines with depth, at all sites, in a negative exponential manner. Yet there is a clear decomposition peak at 8–14 cm. This peak is, therefore, likely to be caused by humification and fermentation processes at this soil depth, sustained by an increased population of decomposing micro-organisms. At the flush sites, there is little indication that the water flow significantly increases average temperature, but it may considerably increase the supply of available nutrients at this depth, possibly stimulating decomposition there.

Both series of strips, inserted in soils at intervals through the summer, present interesting pictures of the interplay between temperature and moisture, in changing the decomposition of the cotton strips. They have also shown depth-related changes in the biological characteristics of the peat and the speed with which decomposing micro-organisms can colonize organic materials, even in the harsh antarctic environment. Clearly, the use of the cotton strip assay has been able to show the patterns of decomposer activity. There are, however, many complex interactions between the decomposition processes and the environmental factors, as indicated by this assay, which still need further evaluation.

7 Acknowledgements
The considerable assistance of D W H Walton is noted with thanks.

8 References

Heal, O.W., Howson, G. French, D.D. & Jeffers, J.N.R. 1974. Decomposition of cotton strips in tundra. In: *Soil organisms and decomposition in tundra,* edited by A.J. Holding, O.W. Heal, S.F. MacLean & P.W. Flanagan, 341-362. Stockholm: Tundra Biome Steering Committee.

Latter, P.M. & Howson, G. 1977. The use of cotton strips to indicate cellulose decomposition in the field. *Pedobiologia,* **17,** 145-155.

Lawson, G.J. 1985. Decomposition and nutrient cycling in *Rostkovia magellanica* from two contrasting bogs on South Georgia. In: *Antarctic nutrient cycles and food webs,* edited by W.R. Siegfried, P.R. Condy & R.M. Laws, 211-220. Berlin: Springer.

Lewis-Smith, R.I.L. 1981. Types of peat and peat-forming vegetation on South Georgia. *Bull. Br. antarct. Surv.,* **53,** 119-139.

Lewis-Smith, R.I.L. & Walton, D.W.H. 1975. South Georgia, Subantarctic. In: *Structure and function of tundra ecosystems,* edited by T. Rosswall & O.W. Heal, 399-423. (Ecol. Bull. no. 20.) Stockholm: Swedish Natural Science Research Council.

Smith, M.J. & Walton, D.W.H. 1985. A statistical analysis of the relationships between viable microbial populations, vegetation and environment in a subantarctic tundra. *Microb. Ecol.,* **11,** 245-257.

Smith, M.J. & Walton, D.W.H. 1988. Patterns of cellulose decomposition in four subantarctic soils. *Polar Biol.* In press.

Walton, D.W.H. 1985. Cellulose decomposition and its relationship to nutrient cycling at South Georgia. In: *Antarctic nutrient cycles and food webs,* edited by W.R. Siegfried, P.R. Condy & R.M. Laws, 192-199. Berlin: Springer.

International comparisons

Use of cotton strip assay in wetland and upland environments — an international perspective

E MALTBY
Department of Geography, University of Exeter, Exeter

1 Summary
Application of the cotton strip assay is examined in a wide range of mainly wetland environments, which include Everglades marsh (USA), Mississippi floodplain (USA), Amazon floodplain (Brazil) and upland moorland (UK). The assay is particularly useful in remote sites, as well as in the spatially and temporally diverse wetland ecosystems.

Cellulose decomposition rates, as indicated by the rate of tensile strength loss of inserted cotton strips, varies between water column and submerged peat in subtropical marsh, but greatest variations in cellulose decomposition rates are caused by differences in nutrient loadings. Results from the Mississippi floodplain indicate marked differences in cellulose decomposition according to flooding regime and pH but, somewhat surprisingly, the most waterlogged anaerobic soil profile with relatively high nutrient status gives the highest cellulose decomposition rates. In the Amazon floodplain, decomposition in an anoxic swamp is as great as an adjacent podzol, and is significantly greater than in an oxisol profile. Cotton strip decay in acid moorland peat increases in response to burning and the application of lime and basic slag.

The cotton strip assay is a potentially powerful method for evaluating cellulose decomposition rates in different wetlands, and, in particular, for investigating the effects of treatment/management alterations or natural changes and trends.

2 Introduction — use in wetlands and uplands
Wetland ecosystems include marsh, swamp, bog, floodplain, shallow lakes and diverse coastal habitats. Water saturation is an important feature of their genesis, but flooding regime, soil and water chemistry, physical properties and vegetation vary greatly in space and time.

Ecological complexity, physical inaccessibility and technical problems of sampling systems at the interface of or between aquatic and terrestrial conditions are reasons why, with the exception of bogs (eg Clymo 1984; Heal *et al.* 1978; Latter *et al.* 1967), very few data exist on organic matter decomposition rates in these systems world-wide. The cotton strip assay has been used at a variety of sites in an attempt to determine the effects of environmental variables and treatments on the organic matter decomposition

cycle, and to produce a range of baseline data on cellulose decomposition in contrasting wetlands. The assay, reviewed thoroughly by Latter and Howson (1977) and by Harrison *et al.* (1988), is favoured for a number of reasons.

i. The methodology is extremely simple and does not rely on sophisticated equipment in the field. It is ideal for use in remote and often extreme environments (in terms of temperature and water-logging), where failure rate with mechanical, electronic or chemical processes is often high.

ii. Insertion, retrieval and preparation for analysis of the strips can be carried out by operators with limited training.

iii. Cotton strips are light and can be airmailed from remote places at relatively low cost, for standardized laboratory analysis.

iv. The flexible uniform cellulose substrate is highly appropriate for investigation of cellulose decomposition across aquatic–soil/sediment interfaces. In fact, one of the possible criticisms of the method in terrestrial soils, ie of the fabric acting as a wick conducting water from the surface of the soil downwards or from wetter lower horizons upwards (Latter & Howson 1977), is of little significance when the soil is saturated or if the strip is actually suspended in a water column.

v. Relatively large numbers of strips can be set out to incorporate the extreme spatial variability and environmental complexity of wetlands.

vi. There is no necessity for continuous monitoring. Periodic sampling can coincide with other field activities in a relatively low-cost experimental design. This point is particularly important in remote sites, where a large proportion of project costs are associated with travel and subsistence.

3 Sites and methods
The full range of wetland or upland sites, conditions and treatments investigated to date are summarized in Table 1. Examples have been selected which demonstrate applications of the technique in detecting environmental influences on decomposition patterns and illustrating variations among ecosystems.

Table 1. Location, generalized sampling details and status of wetland sites where cotton strips were placed

Area	Wetland types, other habitats, soils sampled	Dates	Site data & additional information collected	Collaborators, status of analysis, publication
Everglades marsh Florida, USA (25° 35′ N, 81° 45′ W)	Shark River Slough. Spike-rush/ bladderwort association with standing water and peat substrate. Strips in water column and in upper part of peat profile	1983(pilot) 1984 September– October	Periphyton productivity, macrophytes, nutrients, discharge, height water column, temperature; part of study on effects of increased N and P loading on wetland system	Everglades National Park (G Hendrix), National Park Service (M D Flora). Initial findings in Maltby (1985)
Mississippi floodplain Louisiana, USA (31° 01′ N, 91° 44′ W) (30° 22′ N, 89° 44′ W)	Red River Bay Pearl River Bottomland hardwood forest, floodplain alluvial soils. Sites investigated along a hydrological gradient from non-flooded to frequently flooded. Sampling to 26 cm in 2 cm units	1983(pilot) 1984 August– October	Microbial numbers using dilution plates, chemical and physical soil properties, soil temper-ature, redox, water levels, oxygen; part of study on wetland delineation and differentiation of wetland/ non-wetland soil processes	Laboratory for Wetland Soils and Sediments, Centre for Wetland Resources (Faulkner *et al.* 1987)
Amazon floodplain Brazil, S. America (3° 06′ S, 60° 00′ W)	Sites near Manaus. Swamp (anoxic), creek (high O_2), podzol and oxisol (tropical rainforest) sites. Sampling to 26 cm in 2 cm units	1984– present	Part of pilot study investigating ecological differences between wetland sites	W J Junk, Max Plank Institute. Raw data
Upland moorland North York Moors, UK (54° 24′ N, 0° 58′ W)	Glaisdale and Rosedale Moors. Bog (mainly cotton-grass), thin blanket peat, peaty gley/thin iron pan stagnopodzol complex (mainly heather). Includes sites severely burnt (1976), and those in cycle of normal management rotational burning. Sampling to 26 cm in 2 cm units	1980–83 (all seasons)	Microbial numbers using dilution plates, chemical and physical soil properties; part of study of fire effects on biological properties of organic horizons	Jefferies (1986) Some results available in North York Moors National Park: Moorland Management and Research, 1977–84
Exmoor, UK (51° 10′ N, 3° 50′ W)	Halscombe Allotment and Old Barrow Down. Bog (mainly moor-grass), peat stagnohumic gley/ thin iron pan or stagnopodzol mosaic (acid grassland — heather). Pinkery. Moorland on stagnohumic gley soil reclaimed in 19th century. Sampling depth 10–26 cm depending on profile	1982– ongoing	Microbial numbers using dilution plates, chemical and physical soil properties; part of study on effects of surface fertilizer treatment and grazing on soil properties	Maltby (1986)

Cotton strip data are also available for sites in New Zealand, Canada, Australia, North Carolina, USA, and the Falkland Islands

3.1 Everglades marsh: evaluation of water quality stan-dards on ecological processes — effects of increased nitro-gen (N) and phosphorus (P) loadings on decomposition in the water column and submerged peat

A marshland site was established in April 1983 on one of the main channels, Shark River Slough, through which water flows into the Everglades National Park. The overall aim was to examine the effects of in-creased nutrient loading on water quality, and to deter-mine the specific responses of periphyton (diatom/ algae complex) and macrophytes. The overall objec-tive of the experiment was to assess the adequacy of the present standards for the quality control of water entering the Park (Flora *et al.* 1985). The experimental site is a 'wet prairie', dominated by a spike-rush/blad-derwort (*Eleocharis/Utricularia*) plant association. Three open-ended channels (Plate 11) were con-structed parallel to the water flow and partitioned by fibre-glass sheets extending into the underlying peat to prevent cross-leakage. The centre channel received phosphate-phosphorus and nitrate-nitrogen at a rate of 10 l day^{-1}, with an initial concentration of 3000

Plate 11. *Open-ended channels used for additions of phosphate and nitrate in Shark River Slough in Everglades marsh (Photograph E Maltby)*

mg l^{-1}, from a point source at the upstream end of the channel. The outside channels were maintained as controls until September 1983, when additions of P only to one and N only to the other were initiated. A zone in the marsh immediately outside the N-only channel was maintained as a control zone throughout the entire experimental period, which terminated in autumn 1984. Sampling stations were established 10, 20, 35, 65 and 95 m from the point source of nutrient addition in the test channels, and 20 and 65 m downstream in the control zone. Over the experimental periods, water depth varied generally between 55 cm and 65 cm, and pH was 6.8–8.6 in the control but with diurnal ranges of 7.2–8.5 recorded (D J Scheidt pers. comm.). Detailed patterns of nutrient uptake, water chemistry variations, periphyton productivity and vegetation are given in Flora *et al.* (1985).

The cotton strip assay was used to measure comparative differences in cellulose decomposition induced by the nutrient loadings. In a pilot study in September 1983, only the upper 30 cm of the water column (lowest strip level sampled 27–29 cm) was examined, with strips submerged for 14 days (Maltby 1985). The experiment was repeated in September–October 1984, with 2 sets of 4 strips, 3 to 30 cm depth and one to 55–67 cm (depending on depth to peat substratum) at each sampling station. One set was recovered after 14 days and the other after 28 days' submergence.

3.2 Mississippi floodplain soils: differential effects of flooding regime and soil pH on cellulose decomposition

At Pearl River and Red River bay in the lower Mississippi floodplain, transects were set up within a hardwood forest site to run along a hydrological gradient from dry upland (plot 1) to regularly flooded bottomland sites (plots 4 and 5) (Plate 9). Five soil profiles were examined on the Pearl River transect which had low-nutrient soils with a mean pH of 4.7, and 4 soil profiles were examined along the Red River bay transect which had relatively high-nutrient soils and a mean pH of 7.5.

Cotton strips were inserted in both transects for various time periods from 7–42 days in August–October in 1983 and 1984. At the same sites, soil microorganisms were enumerated, along with a continuous monitoring programme of water levels, moisture content, oxygen, redox, soil chemistry and vegetation, aimed primarily to identify means of delineating wetland and non-wetland areas according to functional attributes (Faulkner *et al.* 1987).

3.3 Amazon floodplain, Brazil: variation in decomposition rates among wetland sites

A range of trial sites close to Manaus were used, and cotton strips were inserted for 14 and 28 days.

  i. anoxic swamp along rainforest creek;
 ii. oxygenated creek (flowing water);
iii. podzol;
 iv. oxisol under tropical rainforest with distinct leaf litter horizon.

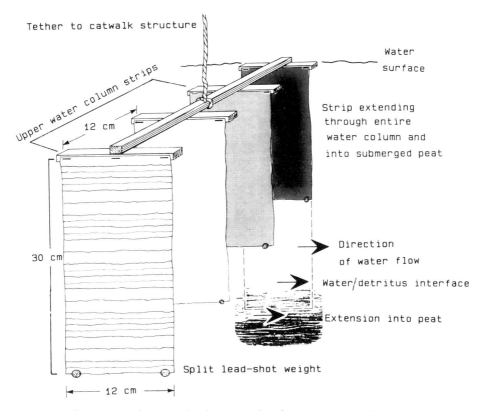

Figure 1. Arrangement of cotton strips attached to wooden frame used to determine decay rates in the water column and submerged peat (as used in Everglades marsh)

3.4 Upland moorlands: effects of management on cellulose decomposition in moorland peat and soils

3.4.1 *North York Moors*

A range of sites on Glaisdale Moor and Rosedale Moor, north Yorkshire, have been investigated to examine the influence of both controlled heather (*Calluna vulgaris*) burning and severe peat fire on decomposition. More details on morphological, chemical and microbiological characteristics of the sites are given in Maltby (1980) and Jefferies (1986). Only the effect of a severe peat fire occurring in 1976 on Rosedale, described in detail in Maltby (1980), is examined here. Cotton strips were inserted in August 1980 for various time periods (6–30 weeks) in 2 areas:

i. where fire in 1976 had removed the entire heather cover of the site and a surface crust had developed on the soil sufficient to deter subsequent plant reestablishment;

ii. the control site unaffected by the 1976 fire and not subjected to normal rotational burning within the previous 10 years.

3.4.2 *Exmoor*

Moorland sites on Exmoor which had management contrasts, eg of fertilizer additions, grazing intensity, or reclamation history, were examined.

3.4.2.1 Two sites were used which had different surface fertilizer treatments: Halscombe allotment, with a comparison of a fertilized, grazed area with an unfertilized, ungrazed area; and Old Barrow Down, with grazed areas with and without fertilizer treatment.

Halscombe allotment is an acid grass moorland (mainly moor-grass (*Molinia*)) site, with a stagnohumic gley soil. The fertilized plot was treated with lime (5 t ha^{-1}) and basic slag (1.5 t ha^{-1}) in October 1978. The unfertilized and ungrazed control plot is a mosaic of grass/heather moorland, dominated by purple moor-grass (*Molinia caerulea*) or by heather. It was fenced to exclude livestock in June 1979. Old Barrow Down is a grass/heather moorland site, with a thin stagno-podzol soil. The fertilized plot was treated with lime and slag in July 1982, and is drier with a greater dominance of heather. The unfertilized plot is open to grazing and is wetter and more peaty. At both sites, 30 cotton strips were buried vertically in the soil in a seasonal series starting in November 1982, with successive insertions in February, May and August. They were retrieved after approximately 3, 6 and 9 months.

3.4.2.2 A grazed site at Pinkery on the southern flanks of the Chains was used as an example of a 19th century reclaimed site. The Pinkery site is moorland originally, with a stagnohumic gley soil. On the reclaimed site, the soil has become modified by subsequent cultivation (burning, ploughing, reseeding and fertilization with lime and basic slag) in 1845–47 (Maltby 1975), and subsequent management to produce a less acidic brown topsoil (pH 5.2–5.5) with a relatively well-drained soil profile. The adjacent unreclaimed site is moor-grass moorland; it is poorly drained with a distinct peaty topsoil and is more acidic (pH 3.8–4.5). Cotton strips were buried at the Pinkery site for 1- and 2-month periods.

Plate 12. Insertion of cotton strips attached to wooden frame in Everglades marsh (Photograph E Maltby)

3.5 Cotton strip assay

Insertion in peats, soils and firm sediments generally followed the technique of Latter and Howson (1977). In the case of water column studies, such as in the Everglades, the strips were stapled to a wooden frame float (4 replicates per frame) and weighted with split lead shot so that the strips hung vertically and the top coincided with the water surface (Figure 1 & Plate 12). The frame accommodates any oscillations in water level over the exposure period, and so maintains a consistent profile with respect to the top of the water column. Multiple frames are used to test cotton strips after different submergence periods or to increase the number of strips per site. In the absence of extreme fluctuations in shallow water columns, strips were then extended into the underlying detritus, soil or sediment, so establishing a continuum from the aquatic to the substratum system. This procedure has been carried out successfully using strips up to 67 cm long in a low velocity flow within the Everglades, and the technique has been extended more recently to experimentally controlled water level cells of a Canadian marsh ecosystem.

The standard strip measured 12 cm x 30 cm. In cases where strips extended from the top of the water column into soil or sediment, the length of strip was increased depending on the water depth. Control and test strips were washed gently in deionized water and air dried before analysis or dispatch to the laboratory. Strips were cut into horizontal substrips 3 cm wide and reduced by fraying to 2 cm, to give test substrips corresponding to depths of 0–2, 3–5, 6–8 cm, etc. Tensile strength (TS) was measured using a tensometer (Monsanto Type W) with 7.5 cm wide jaws set 3 cm apart. All measurements were carried out at 18–22°C and with a relative humidity (RH) of 100% obtained by soaking strips in deionized water.

Individual and mean TS loss of cotton (CTSL) was calculated using the mean field control values obtained for each site (strips inserted and removed immediately). Data were expressed as CTSL % day^{-1} for each level in the profile, and progressive CTSL were plotted against submergence time. The number of strips used varied from 4 in water column studies to 10 in the case of peats and soils.

4 Results and discussion

4.1 Everglades marsh: effects of N and P loadings on cellulose decomposition

Changes in CTSL after 14 days' submergence are shown with respect to (i) the depth in the water column (Figure 2), (ii) distance downstream from the nutrient addition (Table 2, Figures 3 & 4), and (iii) differences between the water column and submerged peat (Table 3). The patterns after 14 and 28 days' submergence are similar, although estimates of cotton rotting rates, calculated as CRR (Hill *et al.* 1988), are generally reduced at 28 days. The 14-day data are preferred because, with increased time of submergence, algal growth on the strips increases the potential for erroneously reduced CTSL.

4.1.1 *Cellulose decomposition in the water column profiles*

A number of salient features emerged (Figure 2):

i. faster cellulose decomposition at almost all depths in the N + P channel, compared with N or P only or the control;
ii. a zone of higher cellulose decomposition in the uppermost 5 cm of the water column in all channels;
iii. reduction in cellulose decomposition at or close to the water/peat interface, especially in the N + P channel;
iv. a sharp rise in cellulose decomposition in the submerged peat of the N + P and P-only channels, less clear in N-only, and absent from the control zone.

4.1.2 *Changes in cellulose decomposition downstream from nutrient addition*

Mean cellulose decomposition rates in the upper 30 cm of the water column differ considerably among

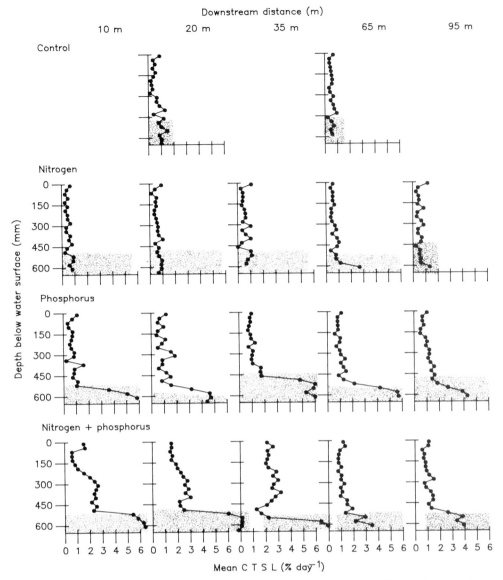

Figure 2. *Everglades marsh: detailed profiles of cellulose decomposition shown by tensile strength loss of cotton strips (CTSL % day⁻¹) in the water column and submerged peat, September–October 1984. Hatching indicates position of submerged peat below water column*

the treatments and with distance downstream from the point of nutrient loading. The N + P treatment produces consistently high values, which peak sharply at 35 m before declining downstream. There is a much reduced response in the P-only channel and virtually no change downstream in the N-only or in the control zone (Figure 3).

The response of cellulose decomposition in the submerged peat is slightly different, with a peak in CTSL in the N + P treatment much closer to the nutrient source, and a considerably sharper change in the P-only channel (Figure 4).

4.1.3 *Water column/submerged peat contrasts*
The ratio of CTSL % day⁻¹ in the upper 10–15 cm of submerged peat to that in the upper 30 cm of the water column provides an index of cellulose decomposition within the detrital substratum in comparison to the water column (Table 3). The ratios confirm that CTSL is consistently higher in the submerged peat, where more anoxic conditions might be expected to prevail,

than in the upper water column, where oxygen levels will be higher but organic substrates for microbial respiration much less available. Differentials between the peat and water column are clearly inflated in the P-only treatment and consistently raised in the N + P channel above values in the N-only and control zone.

Consequences of nutrient enrichment on ecosystem attributes, such as productivity and cellulose de-

Table 3. *Everglades marsh: ratio of tensile strength loss of cotton in peat to upper 300 mm of the water column, September–October 1984, based on 14-day insertion period*

Station downstream distance (m)	Control	P only	N only	N + P
10		5.76	1.53	4.68
20	2.61	5.49	1.59	3.74
35		6.10	1.74	2.78
65	1.17	5.91	2.71	2.95
95		3.18	1.32	3.65

Table 2. Everglades marsh: cellulose decomposition rates (tensile strength loss of cotton strips (CTSL) after 14-day insertion period) and water chemistry of experimental channels, with distance downstream from nutrient addition, September–October 1984

| Station downstream distance (m) | Mean % CTSL day^{-1} (± SE) | | | | Water chemistry (100 mm depth) (g l^{-1}) | | | | | |
| | Upper 300 mm water column | | Submerged peat | | Nitrogen | | | | Phosphate | |
					NO$_3$	NH$_4$	Organic	Total	PO$_4$	Total
Control										
20	0.403	(0.07)	1.050	(0.09)	10	45	1405	1405	8	15
65	0.483	(0.04)	0.563	(0.07)	10	45	1310	1310	7	10
P only										
10	0.673	(0.07)	3.877	(1.05)	10	45	1315	1465	24	30
20	0.831	(0.12)	4.564	(0.15)	10	45	1455	1665	17	50
35	0.919	(0.05)	5.603	(0.25)	10	50	1450	1515	15	35
65	0.875	(0.05)	5.164	(0.48)	10	75	1395	1465	16	25
95	0.826	(0.08)	2.627	(0.50)	10	50	1405	1565	15	30
N + P										
10	1.278	(0.20)	5.984	(0.22)	50	35	1365	1660	19	70
20	1.883	(0.15)	7.044	(0.08)	40	35	1320	1355	14	60
35	2.396	(0.11)	6.659	(0.26)	40	40	1290	1600	15	35
65	0.972	(0.06)	2.863	(0.39)	25	45	1310	1350	11	40
95	0.941	(0.08)	3.433	(0.30)	15	45	1310	1475	11	25
N only										
10	0.449	(0.05)	0.689	(0.13)	75	40	1410	1085	8	20
20	0.490	(0.06)	0.781	(0.06)	105	35	1265	1485	7	17
35	0.516	(0.07)	0.897	(0.10)	80	70	1680	1840	8	22
65	0.496	(0.03)	1.342	(0.45)	80	50	1300	1545	5	15
95	0.549	(0.09)	0.724	(0.15)	75	45	1305	1305	5	15

composition, are not well understood in flowing water systems, especially those in sub-tropical environments. Phosphorus loading consistent with previously agreed water quality standards, yielding 20 mg l^{-1} soluble reactive P (East Everglades Resources Planning Project 1980), has had a major influence on cellulose decomposition rates in the water column, producing an order of magnitude increase in submerged peat. The levels of NO$_3$-N loading are less than 10% of the recommended standard of 700 mg l^{-1}. Although they are 5–6 times the background of the control channel, there is no significant increase in cellulose

decomposition rate in the N-only treatment for either the water column or the submerged peat. In this system at least, N does not seem to be a significant factor limiting aquatic cellulose decomposition. The N + P treatment produces clearly the largest increase in decomposition in the water column, giving a maximum differential 6 times that of the control (Figure 3) and

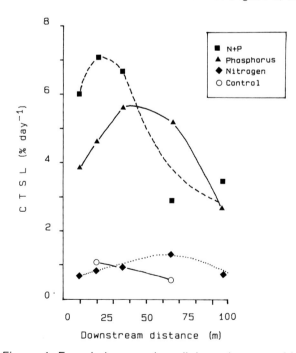

Figure 3. Everglades marsh: cellulose decomposition in upper 300 mm of water column at various distances downstream from point source of nutrient addition, September–October 1984

Figure 4. Everglades marsh, cellulose decomposition in submerged peat at various distances downstream from point source of nutrient addition, September–October 1984

an order of magnitude increase in the submerged peat (Figure 4). Influence of added N appears to raise decay rates in the water column but not in the peat substratum. Overall, it is P which seems to control cellulose decomposition rates. Added P significantly enhances cellulose decomposition by the anaerobic microbial populations at submerged depths in the peat. Where mineral nutrient deficits are rectified by nutrient additions, the submerged peat habitat is clearly more favourable than the water column for decomposer activity.

The pattern of cellulose decomposition with distance from the nutrient source is influenced by low rates of mixing in the slow water flow (17–22 l sec^{-1} in August–October 1984) of this system and progressive

fixation of added nutrients by plants, other organisms and organic matter (Maltby 1985).

The effect of short-term dosage levels, comparable with those which might be permitted to enter the wetland ecosystems of the Everglades National Park, has been a significant increase in cellulose decomposition rates to at least 95 m downstream. This response has major implications for ecological processes and system stability. Such major changes in cellulose decomposition could not be predicted directly from water chemistry data. Yet the cotton strip assay has provided a clear picture of the nutrient impacts and dispersal in both the water column and submerged peat simultaneously. Further experimental work is proceeding to explain some of the detailed

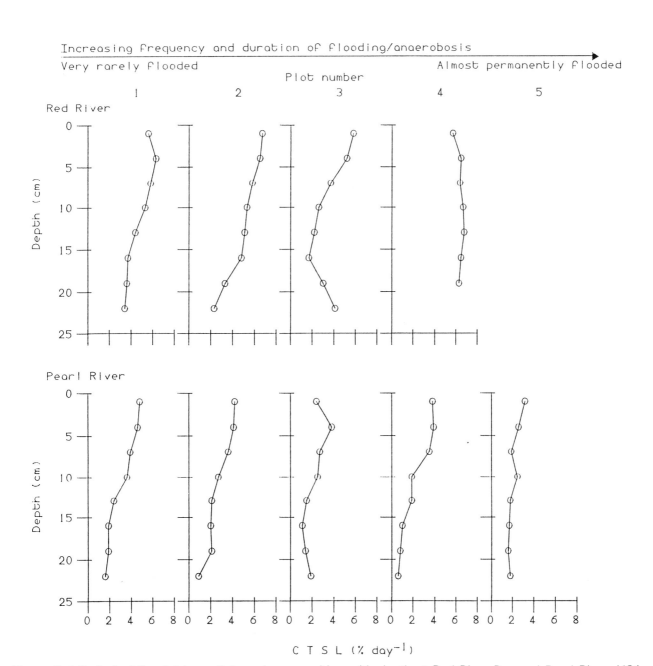

Figure 5. Mississippi floodplain: cellulose decomposition with depth at Red River Bay and Pearl River, USA. Plot numbers represent hydrological gradient from dry (1) to wet (4 and 5)

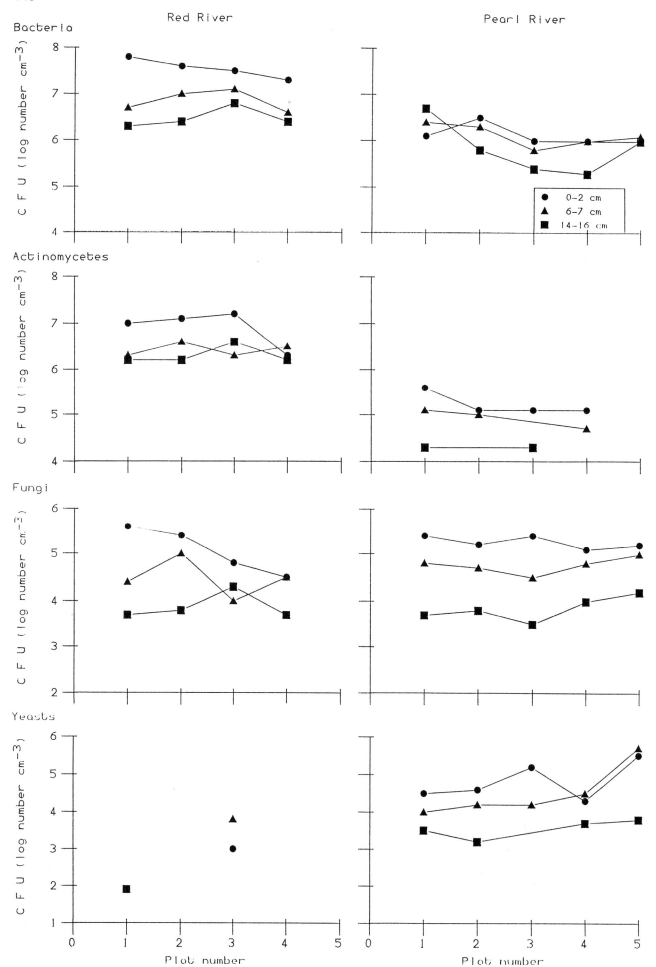

Figure 6. Mississippi floodplain: numbers (colony-forming units) of bacteria and actinomycetes on cellulose media (Faulkner et al. 1987) and filamentous fungi and yeasts on Martin's media for Red River and Pearl River profiles

patterns of decay, with particular reference to the fate of nutrients in the peat detritus and different periods of strip insertion. The technique offers scientists and wetland managers an inexpensive, versatile and relatively quick technique for detecting ecological impacts, which can be used in the planning of buffer zones for water quality control and the maintenance of ecological stability. In the Everglades experiment, the results indicate that adherence to present standards will not prevent significant changes occurring in the natural ecosystem due to more nutrient-rich waters entering the Park from its increasingly more farmed periphery.

4.2 Mississippi floodplain: differential effects of flooding regime and soil pH

For Red River Bay and Pearl River, CTSL % day^{-1} is plotted against soil profile depth for 2 weeks' insertion in 1984 (Figure 5). Cellulose decomposition is generally much higher in the more nutrient-rich, high pH Red River soils than in the nutrient-poor, low pH Pearl River soils. This result relates well to the generally higher numbers of bacteria and actinomycetes enumerated on cellulose media at the Red River sites (Figure 6). However, yeast numbers are higher at Pearl River, and there are no major differences in filamentous fungi estimates between the 2 areas, except for Red River Bay at the wet end of the transect where numbers generally declined (Figure 6).

The differentiation along the hydrological gradient, particularly marked in Red River Bay, showed surprisingly that the overall CTSL throughout the profile is highest in the most waterlogged anaerobic plot (plot 4) and least in one of the intermediate soil profiles (plot 3) which is periodically stressed either by flooding or by drying out. Differences among plots are least at

the surface and greater in the more anaerobic zones of the soil profiles below 14 cm (Figure 5). The anaerobic profile (plot 4) is characterized by high CTSL which is maintained with depth, in contrast with the more aerobic profiles, which have high surface cotton decay rates that decline progressively with depth (eg plots 1 and 2). In between these extremes of decay rate, the intermediate soil profile at plot 3 gives a sharp increase in CTSL at depth where the soil environment is more constantly anaerobic and the microbiota more adapted to anaerobic decomposition. This result implies that some soils may have microflora well adjusted to cellulose breakdown under anaerobic conditions, some under aerobic conditions, but others where neither condition is constant may lack close adjustment to either. Such differentiation is less clear in the low pH environment at Pearl River, and may explain why it has not been readily observed in previous studies of cotton strip decay in anaerobic systems which have been concentrated in very acid, generally peat-forming environments. The Mississippi floodplain studies reported in more detail elsewhere (Faulkner *et al.* 1987) underline that cellulose breakdown may not necessarily be inhibited by anaerobic conditions *per se*, and that the cotton strip assay may be a most valuable tool differentiating wetland/non-wetland soil regimes on the basis of intrinsic soil processes.

4.3 Amazon floodplain: variation in cellulose decomposition rates

Decomposition profiles derived from the 28-day insertion time are given in Figure 7. Tukey's Honestly Significant Difference test showed the differences between sites to be oxisol>swamp/podzol at P<0.001 (D M Howard pers. comm.). The creek site is excluded from statistical analysis because of a lack of sequential depth values. Overall, 14 days' exposure

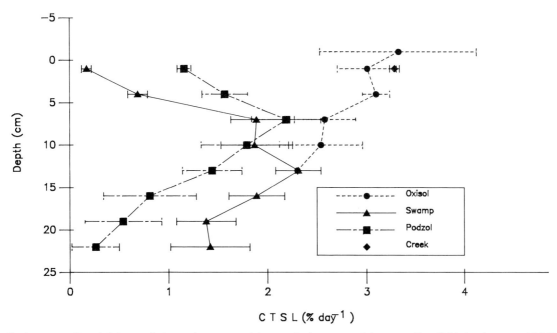

Figure 7. Amazon floodplain: cellulose decomposition at 4 sites near Manaus, Brazil (July–August 1984): oxisol, anoxic swamp, podzol and aerobic creek, mean of 5 ± SE

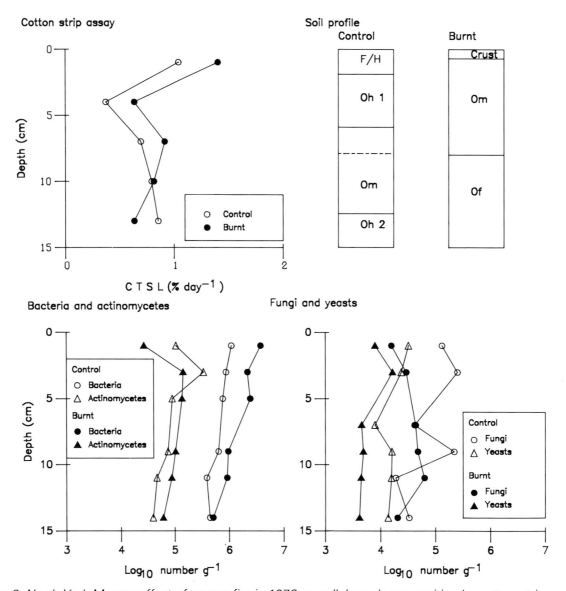

Figure 8. North York Moors: effect of severe fire in 1976 on cellulose decomposition by cotton strip assay and microbial numbers with depth in peaty horizon of stagnohumic gley (determinations by J Jefferies 1981) (standard errors for CTSL and microbial numbers were all within the points). The broken line in control indicates a structural break. Soil profile terms in Figures 8 and 10 according to Avery (1980)

values are greater than those for 28 days (P<0.001), but this fact does not alter the relationship of sites, and there is no significant interaction between sites and exposure time.

4.4 Upland moorlands: effect of management on cellulose decomposition in peat and soils
4.4.1 *North York Moors*
Comparative results of cotton strip assay after 6 weeks' insertion and dilution plate counts of microorganisms are given for severely burnt and control peat profiles in Figure 7. There is a clear increase in CTSL due to fire in the upper part of the soil profile, and particularly at 3–5 cm depth. The increase in cotton decay rate corroborates earlier findings of a surge in microbial numbers (Figure 8; Maltby 1980), which results in increased biological activity in the peat for at least 5 years after the fire.

'Normal' rotational prescribed burning also stimulates cotton strip decay rates (Jefferies 1986), and evidence

discussed elsewhere suggests that the peat cover on the North York Moors is generally becoming progressively more aerobic (Maltby 1980, 1986b). The cotton strip assay will provide a useful method for the comparative assessment of cellulose decomposition under different burning intensities and strategies valuable in the development of sound moorland management techniques.

4.4.2 *Exmoor*
4.4.2.1 Seasonal variations of cotton strip assay rates in soil profiles receiving a surface treatment of fertilizer and in untreated control areas are shown in Figure 9, using insertion periods of approximately 3 months. Results indicate the sharp fluctuations in CTSL according to season and between years, the variation being particularly marked at lower sampling depths (eg 12–14 cm). A retrieval problem encountered on the surface-treated sites during summer 1984 and 1985 was that high rates of decay, probably as a result of the particularly wet summers, caused detachment of the

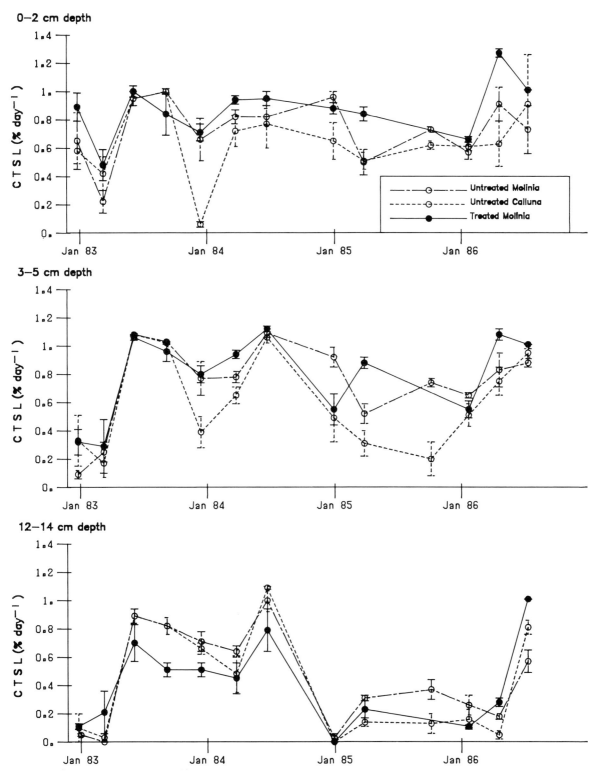

Figure 9. Exmoor, Halscombe allotment: fertilizer treatment of a stagnohumic gley. Variation in cellulose decomposition with time for the peat horizon. The sites were areas dominated by moor-grass (Molinia) with surface treatment of lime/basic slag applied in 1978; untreated moorland, fenced, dominated by moor-grass; untreated moorland, fenced, dominated by heather (Calluna). Means ± SE

tops of some strips, making relocation impossible. Before 1984, there was no evidence of consistently higher decomposition rates in the surface-treatment sites compared with the control. Since then, the 2 treated sites, at Halscombe and Old Barrow Down, have shown higher rates of cellulose decomposition at the surface (down to 3–5 cm by the end of 1985). There is evidence for the effect extending to 12–14 cm at Halscombe by the spring of 1986, indicative of a progressive change in soil biological activity extending over time down the soil profile. The main flush of microbial activity due to fertilizer application lasted only until 1981 (Maltby 1986a), yet it seems that impact on soil biological processes may well extend far beyond that immediate influence. The extent of further changes and of differentiation between treated

152

and untreated sites will undoubtedly influence planning recommendations for management of the Exmoor uplands (Curtis & Maltby 1980).

4.4.2.2 Differential development of soil profiles due to 19th century reclamation results in strong contrasts

in both soil biological properties and pedogenesis (Maltby 1975, 1977, 1984). There is a marked difference in CTSL in the soil profiles of reclaimed and unreclaimed sites (Figure 10), which is related to contrasts in dilution plate counts of micro-organisms, soil profile morphology and soil chemistry. Higher rates of

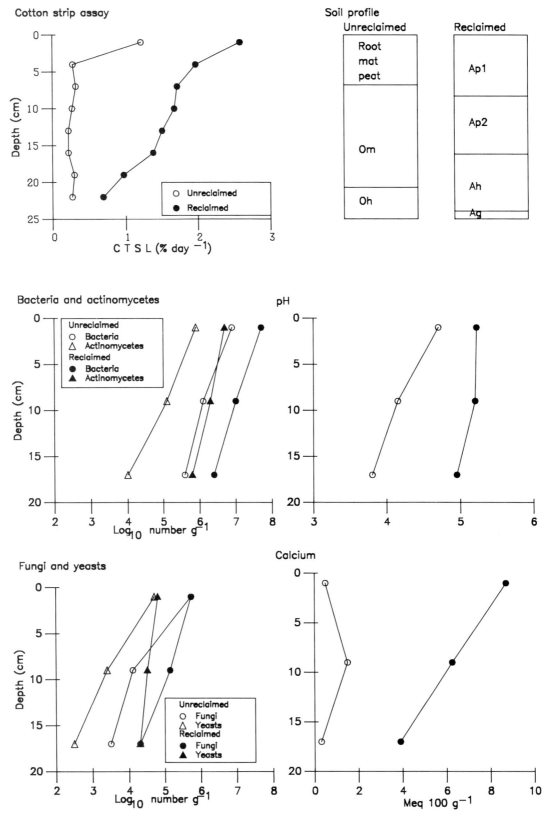

Figure 10. Exmoor, Pinkery: cellulose decomposition and microbial numbers with depth in winter 1984–85 for unreclaimed stagnohumic gley and adjacent profile reclaimed 1845–47. Numbers (CFU) of actinomycetes and bacteria (modified Jensen's media) and filamentous fungi and yeasts (Martin's media); sticky point pH and extractable Ca (determination by L White)

cellulose decomposition within the reclaimed soil help explain the maintenance of a brown topsoil and failure of a peaty horizon to redevelop, despite location in an active peat-forming environment (Maltby & Crabtree 1976).

5 Precautions in application of assay
Experience has underlined several areas requiring particular attention in experimental designs.

5.1 Cotton strips can be inserted by a spade or other tools in soils with only shallow surface flooding, but accurate insertion becomes difficult when water columns exceed 50 cm. In all flooded sites, the water/soil interface can be indicated on the strip with a short cut (which can be made underwater after insertion), and great care must be taken to locate the site with flagging tape in order to aid retrieval. At the other extreme, sites can dry out for parts of the year. In clay-rich profiles, particularly frequent on floodplains, insertion can be impaired and only very poor soil contact achieved due to temporary hardening and shrinkage of the soil. The sampling strategy should be designed to avoid, if possible, the need to insert strips at these times.

5.2 Lack of previous information on decay rates in many environments demands pilot studies to establish the timing of assay periods. This requirement has been particularly important in water column studies, and in experiments in the sub-tropics and tropics. There may still be problems in comparative studies in temperate areas, where strips in some treatments may decompose much more rapidly than in control sites in certain seasons (eg Exmoor).

5.3 Decay is so rapid in sub-tropical and tropical wetland environments that any all-year-round investigation using the Shirley Soil Burial Test Fabric would be impracticable, unless carried out locally, and, in any case, will be much more time-consuming than in temperate or high-latitude regions (a heavier form of cloth could, however, be used).

5.4 Biotic interference with strips can be a periodic problem in wetland or marginal wetland sites. This interference includes algal growth, which may induce some TS increase due to filamentous binding, and root growth which may penetrate long-exposed strips. It is generally necessary to discard strips subjected to this degree of disturbance. Fauna, especially termites and crayfish in low-latitude floodplains and swamps, may eat through strips, and in grazed sites sheep or other herbivores may pull up the strips or eat off the exposed tops. Vickery and Floate (1988) used mesh cages to deter sheep.

6 Conclusion
The cotton strip assay offers a surrogate and averaging measure of detailed and complex biological processes in soil, sediment and aquatic environments. It is poten-

tially powerful in differentiating a wide range of ecological environments and in measuring the comparative effects of treatments or natural changes and trends. The assay integrates the influence of many factors, only some of which have been illustrated here. Considerable research effort must still be directed towards explaining decay rates and interpreting patterns in the light of ecological or pedogenic data. Recent efforts to standardize data using a linearization approach (Hill et al. 1988) are an encouraging basis for further work. The modelling of decay rates in terms of climatic and other variables (Ineson et al. 1988) provides insight to understanding global and environmental patterns. However, the greatest value of the cotton strip methodology will probably remain primarily as a means of comparative analysis within individual studies. The assay offers scope for considerable development and improvement in understanding the behaviour, reaction and management of extremely complex ecosystems in future research on decomposition and related processes in wetlands.

7 Acknowledgements
Mark Flora, David Walker, Rosemary Maltby and Dan Scheidt are thanked for their help in the Everglades. Mark Flora kindly made available unpublished water chemistry data, and the study was made possible by financial assistance from Everglades National Park. Dr Gary Hendrix, Research Director, is thanked for his particular support. Steve Faulkner assisted in the Louisiana study, which has been supported by a research contract from the US Army Corps of Engineers, Waterways Experiment Station, Vicksbury, MS. Jennette Jefferies carried out the North York Moors analyses, and Linda White the sampling of Exmoor unreclaimed and reclaimed profiles. Investigation of the effects of surface treatment and grazing at Halscombe and Old Barrow Down has been supported financially by the Exmoor National Park. Dr Wolfgang Junk is thanked for his help in sampling the Amazon floodplain. The statistical analyses of the Amazon data were carried out at the Institute of Terrestrial Ecology's Merlewood Research Station.

8 References
Avery, B.W. 1980. *Soil classification for England and Wales.* (Soil Survey technical monograph no. 14.) Harpenden: Soil Survey of England and Wales.
Clymo, R.S. 1984. The limits to peat bog growth. *Phil. Trans. R. Soc. B*, **303**, 605-654.
Curtis, L. & Maltby, E. 1980. Conserve or concede. How best to strike the balance of use in Britain's smallest National Park — Exmoor. *Fmrs' Wkly,* 27 June.
East Everglades Resources Planning Project. 1980. *Proposed management plan for the East Everglades.* Miami: Metropolitan Dade County Planning Dept.
Faulkner, S.P., Patrick, W.H. Jr., Maltby, E., Gambrell, R.P. & Parker, W.B. 1987. *Characterisation of bottomland hardwood wetland transition zones in the lower Mississippi river valley.* Report of U.S. Army Corps. of Engineers, Waterways Experiment Station, Vicksburg.

Flora, M.D., Walker, D.R., Scheidt, D.J., Rice, R.G. & Lander, D.H. 1985. *The response of experimental channels in Everglades National Park to increased nitrogen and phosphorus loading: Part 1 — Nutrient uptake and periphyton productivity.* South Florida Research Centre Report. (Unpublished.)

Harrison, A.F., Latter, P.M. & Walton, D.W.H., eds. 1987. *Cotton strip assay: an index of decomposition in soils.* (ITE symposium no. 24.) Grange-over-Sands: Institute of Terrestrial Ecology.

Heal, O.W., Latter, P.M. & Howson, G. 1978. A study of the rates of decomposition of organic matter. In: *Production ecology of British moors and montane grasslands,* edited by O.W. Heal & D.F. Perkins, 136-159. Berlin: Springer.

Hill, M.O., Latter, P.M. & Bancroft, G. 1988. Standardization of rotting rates by a linearizing transformation. In: *Cotton strip assay: an index of decomposition in soils,* edited by A.F. Harrison, P.M. Latter & D.W.H. Walton, 21-24. (ITE symposium no. 24.) Grange-over-Sands: Institute of Terrestrial Ecology.

Ineson, P., Bacon, P.J. & Lindley, D.K. 1988. Decomposition of cotton strips in soil: analysis of the world data set. In: *Cotton strip assay: an index of decomposition in soils,* edited by A.F. Harrison, P.M. Latter & D.W.H. Walton, 155-165. (ITE symposium no. 24.) Grange-over-Sands: Institute of Terrestrial Ecology.

Jeffries, J. 1986. *The effect of burning on selected biological and physico-chemical properties of surface peat horizons on the North York Moors.* PhD thesis, University of Exeter.

Latter, P.M. & Howson, G. 1977. The use of cotton strips to indicate cellulose decomposition in the field. *Pedobiologia,* **17,** 145-155.

Latter, P.M., Cragg, J.B. & Heal, O.W. 1967. Comparative studies on the microbiology of four moorland soils in the Northern Pennines. *J. Ecol.,* **55,** 445-464.

Maltby, E. 1975. Numbers of soil micro-organisms as ecological indicators of changes resulting from moorland reclamation on Exmoor, UK. *J. Biogeogr.,* **2,** 117-136.

Maltby, E. 1977. *Ecological indicators of changes in soil conditions resulting from moorland reclamation on Exmoor.* PhD thesis, University of Bristol.

Maltby, E. 1980. The impact of severe fire on *Calluna* moorland in the North York Moors. *Bull. Ecol.,* **11,** 683-708.

Maltby, E. 1984. Response of the soil microflora to moorland reclamation for improved agriculture. *Pl. Soil,* **76,** 183-193.

Maltby, E. 1985. Effects of nutrient loadings on decomposition profiles in the water column and submerged peat in the Everglades. In: *Tropical peat resources — prospects and potential. Proc. int. Peat Soc. Symp.,* 450-464.

Maltby, E. 1986a. *Soil property responses to surface treatment and associated management practices at Halscombe Allotment/ Humbers Ball and Old Barrow Down/Hawkridge Plain 1978–86.* Exmoor National Park Authority.

Maltby, E. 1986b. Investigations of the peat cover of the North York Moors. In: *Moorland management,* edited by D.C. Statham, 28-41. Helmsley, Yorkshire: North York Moors National Park Authority.

Maltby, E. & Crabtree, K. 1976. Soil organic matter and peat accumulation on Exmoor: a contemporary and palaeoenvironmental evaluation. *Trans. Inst. Br. Geogr., N.S.* **1,** 259-278.

Vickery, P.J. & Floate, M.J.S. 1988. Effects of lime and pasture improvement on cotton strip decomposition in three Scottish acid hill soils. In: *Cotton strip assay: an index of decomposition in soils,* edited by A.F. Harrison, P.M. Latter & D.W.H. Walton, 109-112. (ITE symposium no. 24.) Grange-over-Sands: Institute of Terrestrial Ecology.

Decomposition of cotton strips in soil: analysis of the world data set

P INESON, P J BACON and D K LINDLEY
Institute of Terrestrial Ecology, Merlewood Research Station, Grange-over-Sands

1 Summary
This paper reports the results from the statistical analyses performed on the cotton strip data set, obtained during the Symposium workshop on the use of cotton strip assay in decomposition studies. The data set, derived from various workers, contained information from 329 replicated cotton strip insertions. The data were examined using multiple regression, in an attempt to determine the factors which appeared to control the rate of decomposition of cotton strips in soil in a world-wide comparison.

Analysis of the large data set highlighted the problems associated with the limited geographical distribution of strip placement. Analysis of a subset of sites for which an increased number of variables were available revealed the importance of climatic factors in determining cotton strip decay rate. In particular, potential evaporation derived from literature values related very closely to decomposition rate.

2 Introduction
Many attempts have been made by soil ecologists to utilize standard substrates in organic matter decomposition studies, with the expectation that such substrates would enable inter-site comparisons of decomposition to be made, by removing the confounding effect of substrate quality (Swift *et al.* 1979). Substrates have ranged from shoe-laces (Rosswall 1974) through to matchsticks (Abrahamsen *et al.* 1975), with cellulose being the most widely used.

The cotton strip assay has been applied in a wide variety of situations around the world, both to compare effects of various treatments on rates of decomposition at a single site (eg Howson 1988) or at different sites (Heal *et al.* 1974). The inter-site study of Heal *et al.* (1974) compared tensile strength loss of cotton strips (CTSL) at a number of sites in the tundra biome and, although cotton strips have since been used more widely, that remains the only global survey of rates of decomposition using cotton strips.

Global comparisons and models of decomposition processes have been attempted by few workers (Esser *et al.* 1982; Berg *et al.* 1984), and such studies have been aggravated by the paucity of data for comparable substrates at sufficiently widespread locations. Therefore, the current Symposium was seen as an opportunity to collate the existing cotton strip data in order to determine those environmental factors controlling decomposition of this material at the global scale.

During the course of the Symposium, 2 workshop sessions were held in which interactive analyses of the available data were attempted. Suggestions were made by participants as to which analyses should be performed and, from these sessions, there arose several conclusions and suggestions for further work. The current paper is principally a report of the outcome of these workshop sessions and of subsequent re-analysis of the world-wide data set after additional information and amendments had been provided by the relevant authors.

3 Methods
3.1 Original data set
Delegates to the Symposium were asked, prior to the meeting, to provide data on the tensile strength loss following insertion of cotton strips into soils. The variables requested, together with the units of measurement, are outlined in Table 1. The data were collated into a standard tabular form, and then coded in a manner suitable for analysis using the GENSTAT statistical computer package (Alvey *et al.* 1983). All computations were carried out on a Honeywell 66/DPS-300, in batch mode.

A total of 329 cases was included in the full data set, with each case representing the mean value of replicated strips for the same site and treatment. The calculation of the standard cotton rotting rate (CRR), as described by Hill *et al.* (1985), was performed for each case to permit comparison of results, expressed on a yearly basis. Data sets outside the recommended limits of 10–90% tensile strength loss were rejected.

The mean, minimum and maximum values for each of the variables requested from the contributors (Table 2) revealed that several important variables were missing from the data set, either because they had not been measured, or because authors had failed to provide them. Unfortunately, no data were provided for assay period air temperature, soil temperature or soil moisture content for any of the cases, and the majority of contributors were unable to supply data for many aspects of soil chemistry.

3.2 Augmented data set
After the Symposium, the data sets were checked by authors and some further information was provided. It was also decided that estimates for certain variables could be obtained from published sources in order to explain the variance observed in CRR rates for as wide a range of sites as possible, in terms of climatic factors.

Table 1. The variables requested for the full data set

Mnemonic	Variable	Units
Details of strip placement		
DEPTH	Depth in soil	cm
TS	Mean tensile strength	kg
SE	Standard error of the mean of tensile strength	
FC	Field control tensile strength	kg
CLOC	Cloth control	
N	Number of samples	
WIDTH	Frayed width of tested substrip	cm
DAYS	Number of days in the field	
DAYNO	Standard day number when samples were placed	
YEAR	Year	
CLOTH	Cloth batch (colour code)	*
STRIP	Code number for strip	
NAME	Name of worker	*
Site characteristics		
SITE	Name of site	*
PLOT	Name of plot	*
SUBPLOT	Name of subplot	*
COUNTRY	Country	*
LAT	Latitude	
LONG	Longitude	
HABIT	Habitat type	*
VEG	Vegetation type	*
MAN	Form of management	*
ALT	Altitude	m
Climatic variables		
CLIM	Climatic zone	*
TEMP	Mean annual temperature	°C
RAIN	Total annual rainfall	mm
SOILT	Mean annual soil temperature	°C
SOILM	Mean soil moisture	% moist weight
PTEMP	Period mean temperature during strip insertion	°C
PRAIN	Period rainfall	mm d^{-1}
PSOILT	Period soil temperature	°C
PSOILM	Period soil moisture	% moist weight
Soil characteristics		
STYPE	Soil type	*
LOI	Loss on ignition	%
TOTN	Total soil nitrogen	%
EXTN	Extractable soil nitrogen	µg g^{-1}
TOTP	Total soil phosphorus	µg g^{-1}
EXTP	Extractable soil phosphorus	µg g^{-1}
CA	Total soil calcium	µg g^{-1}
EXTK	Extractable soil potassium	µg g^{-1}
PH	Soil pH	

* denotes an alphanumeric string

Values of mean daily temperature, precipitation and potential evaporation calculated for the period of strip insertion were derived from the climatic compilation of Müller (1982), using data for the nearest climatic station at comparable altitude. Where possible, actual climatic data provided by the individual workers were used in preference to those derived from Müller (1982), and there was generally good agreement between these values where both were available.

Table 2. Mean, minimum and maximum values for numerical data in the full data set. The mnemonics used to describe the variables are outlined in Table 1

Variable	Mean	Minimum	Maximum	Number of missing values
Details of strip placement				
DEPTH	8.65	1.50	18.00	0
TS	20.65	3.10	52.00	0
SE	1.94	0.13	5.54	0
FC	36.23	12.00	56.00	0
CLOC	*	*	*	329
N	10.72	4.00	20.00	0
WIDTH	4.02	3.00	5.00	0
DAYS	191.83	14.00	383.00	0
DAYNO	205.70	8.00	341.00	0
YEAR	1975.44	1968.00	1984.00	0
Site characteristics				
LAT	37.56	−65.25	71.28	0
LONG	−6.8	−113.00	156.68	0
ALT	396.09	5.00	1320.00	0
Climatic variables				
TEMP	5.54	−12.50	27.50	38
RAIN	979.91	108.00	2011.00	38
SOILT	34.32	23.00	48.00	190
SOILM	64.70	12.00	99.00	142
PTEMP	*	*	*	329
PRAIN	102.54	92.00	112.00	236
PSOILT	*	*	*	329
PSOILM	*	*	*	329
Soil characteristics				
LOI	57.97	1.00	98.50	37
TOTN	1.56	0.10	3.10	96
EXTN	431.50	2.00	1000.00	285
TOTP	530.59	1.00	1500.00	243
EXTP	66.21	4.00	220.00	69
CA	1561.04	68.00	7906.00	51
EXTK	344.66	1.40	1040.00	51
PH	4.49	3.00	9.20	95

*denotes missing values

The augmented data set was restricted to data for decomposition rates of strips in the top 2 cm of the soil, and to strips not receiving any additional form of treatment. Duplicate site cases were only used if they represented results from insertions during different periods of the year, and this resulted in the selection of 48 cases, representing all of the sites identified in Figure 1.

4 Results of analyses
4.1 Original data set
The analyses of the basic data set are described in 3 sections: (i) a brief discussion of the choice of parameters used to describe the rate of cotton strip rotting; (ii) a brief description of some sample analyses performed during the workshop; and (iii) a more comprehensive investigation of the data set undertaken after the workshop, with the benefit of experience of the analyses done during the workshop. As a result of this latter analysis, several difficulties in analysing

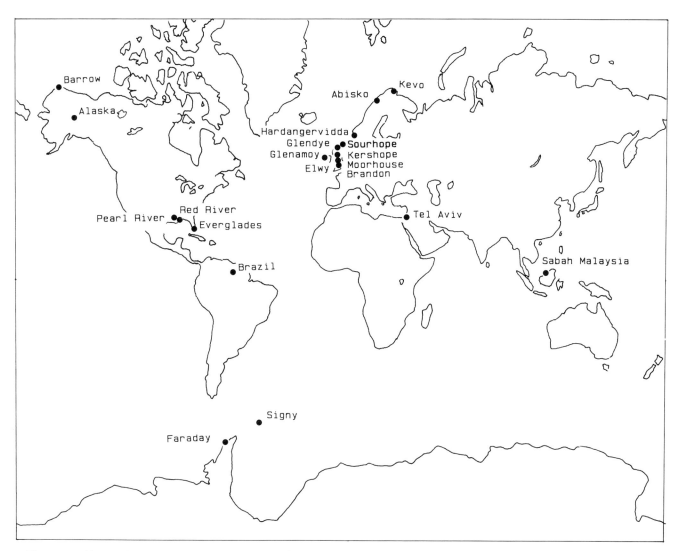

Figure 1. Sites where the cotton strip assay for decomposition rate has been used (●)

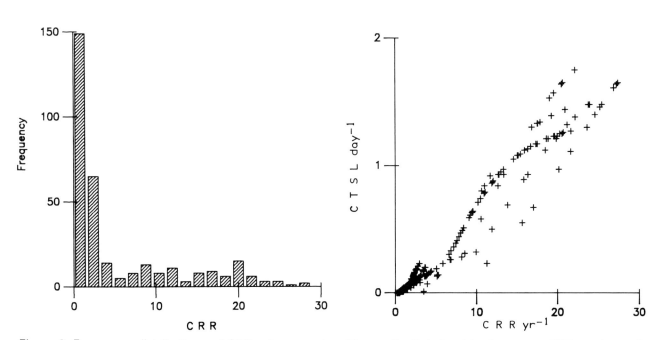

Figure 2. Frequency distributions of CRR values used in the original data set

Figure 3. Relationship between CRR and tensile strength loss (CTSL) day^{-1} for the original data set

the original data set were revealed. The shortness of the section describing analyses requested by delegates during the workshop is largely due to our subsequent realization that the data set was often incompatible with the analyses which had been requested.

4.1.1 *Measurement of cotton strip decomposition*
The work of Hill *et al.* (1985, 1988) has established a 'cotton rotting rate' parameter (CRR), which was specifically designed to be less affected by the known technical difficulties of the procedure. In general, this parameter was used as the dependent variable, describing cotton rotting rates, throughout the analyses. A histogram of its distribution is given in Figure 2, and can be seen to be highly skewed, most values being of small magnitude from 0 to 5 (due to the predominance of cold sites), but with a very long tail to the right. During the workshop, one of the delegates requested the calculation of an earlier measure of cotton rotting rate, CTSL day^{-1}, and, having calculated this measure, we decided, for interest's sake, to see how closely it was related to the CRR parameter of Hill *et al.* (1985). The scatter diagram shown in Figure 3 reveals a surprisingly tight relationship. Indeed, the linear correlation between the 2 variables explains 96% of their variation, but the scatter plot reveals a sigmoid 'backbone' arising from the transformation equation applied by Hill *et al.* (1985), and suggests that the linearization of the decay curve is not fully achieved.

Table 3. CRR as a function of temperature, longitude, absolute latitude and the number of days of strip insertion

Regression coefficients: Y-variate = CRR

	Estimate	SE	t
Constant	30.82	3.39	9.1
TEMP	−0.49	0.13	−3.8
DAYS	−0.03	0.00	−9.3
ALAT	−0.31	0.04	−7.2
LONG	−0.02	0.01	−2.3

Analysis of variance

	Degrees of freedom	Sum of squares	Mean square
Regression	4	11299	2824.68
Residual	323	4972	15.39
Total	327	16271	49.76

Variation accounted for 69.1%

Correlation matrix: df = 326

	CRR	CTSL day^{-1}	TEMP	LONG	ALAT	DAYS
CRR	1.00					
CTSL day^{-1}	0.98	1.00				
TEMP	0.78	0.74	1.00			
LONG	−0.61	−0.57	−0.83	1.00		
ALAT	−0.75	−0.71	−0.94	0.80	1.00	
DAYS	−0.74	−0.71	−0.80	0.51	0.62	1.00

It is somewhat surprising that the less robust conventional measure, CTSL day^{-1}, is so highly correlated with the CRR measure that the 2 could almost be regarded as equivalent. In practice, given that the quality of the cloth is carefully controlled to give a standard tensile strength, and that all investigators endeavour to remove their cloth at a time when it is expected to be around 50% rotted, the correlation is perhaps rather less surprising with hindsight. It does, however, indicate that the procedural difficulties may, in practice, be rather less than were feared. In view of the fact that the CRR measure of Hill *et al.* (1985) is biologically sound and mathematically robust, we have preferred this measure in our subsequent analyses. However, for the benefit of investigators who have used the other method, or those who wish to compare previous results with the values of CRR given in this paper, the regression equation predicting CRR from CTSL day^{-1} was found to be:

$$CRR = 0.8433 + 14.5620 \times CTSL\ day^{-1}$$

4.1.2 *Examples of analyses requested during the workshop*
Several requests were made to investigate the relationship between CRR and fundamental environmental parameters, such as temperature and latitude. The example given below analyses CRR as a function of temperature, longitude, absolute latitude and the number of days the strips were in the soil. These 4 variables were derived from a step-wise regression. Absolute latitude (ALAT) is the latitude expressed as a positive number, representing degrees from the equator, regardless of whether in the northern or southern hemisphere. The analysis is given in Table 3, from which it can be seen that all 4 predictor variables are significantly related, and that the analysis is based on almost the entire data set (327 out of 329).

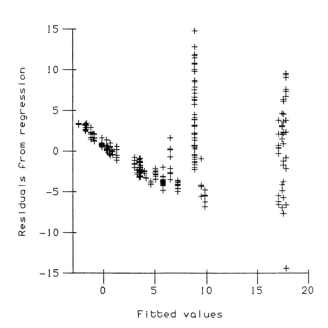

Figure 4. Scatter plot of residuals as a function of fitted values for the regression outlined in Table 3

While the relationship with temperature and latitude, which also has a temperature component, would be expected, the significant effect of the parameter for days is interesting, as the CRR variable would be expected to have removed the time component. Again, this relates to the way in which investigators endeavour to attain 50% CTSL, and reflects their *a priori* judgement of the time required to reach this rate. If we examine the scatter plot of the residuals against the fitted values, given in Figure 4, we see that the assumptions of the regression analysis have not been properly met: there is a broad trend for the residuals to decrease with the magnitude of the fitted values, while 2 (or more) data sets from particular regions give broad scatters of residuals for very similar magnitudes of the fitted values (these appear as vertical scatters of points at the centre and at the right of the diagram of Figure 4). We should further note, when examining the correlation matrix, that the regressor variables are themselves highly interdependent, a fact which makes the use of 'step-up' or 'step-down' procedures of multiple regression highly dubious to identify the better predictor variables. The reason is that multiple regression makes a largely arbitrary choice as to which of a set of highly interdependent variables to include, depending on minor and insignificant differences in the structure of the correlation matrix.

A second analysis was requested, directed more biologically at understanding the process of cotton decomposition, and this analysis examined the relationship between the $\log_e$ CRR and temperature, subsequently adding the $\log_e$ carbon/nitrogen ratio (LGCN) and pH. The scatter diagram of $\log_e$ CRR on temperature is given in Figure 5, from which a broad relationship is evident, explaining 40% of the variation (Table 4). In the full model (CRR = f(temperature, LGCN, pH)), the effect of pH was not significant, but the effects of both temperature and LGCN were as indicated in Table 5, and together explained 43% of the variance in CRR.

4.1.3 *Examination of the original data set*
Straightforward interpretation of multiple regression analyses requires the regressor, or predictor, variables to be uncorrelated and the data set to have multivariate normal distribution. We have already indicated above that many of the variables in the present data set are inter-correlated, which makes interpretation difficult. Furthermore, many of the biologically more interesting variables were not available from many locations represented by this data set. Accordingly, analyses which include the more interesting parameters are perforce restricted to a subset of the data which will often not be representative. We illustrate this difficulty by presenting 2 correlation matrices as Table 6. The full data set comprises 329 sites and, if we confine our interest to 7 basic variables, we are able to utilize information from 236 of these; adding 4 commonly recorded, but not especially interesting, biological variables only reduces the number of sites to 231, but

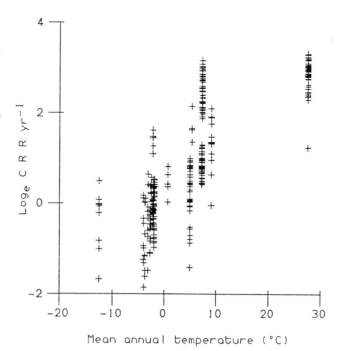

Figure 5. Scatter plot of $\log_e$ CRR against mean annual temperature for the original data set

Table 4. Relationship between $\log_e$ CRR and mean annual temperature

Regression coefficients: Y-variate = $\log_e$ CRR

	Estimate	SE	t
Constant	0.08	0.04	2.1
TEMP	0.09	0.01	12.4

Analysis of variance

	Degrees of freedom	Sum of squares	Mean square
Regression	1	59.60	59.60
Residual	233	89.81	0.38
Total	234	149.41	0.64

Variation accounted for 39.6%

Table 5. Relationship between $\log_e$ CRR, mean annual temperature and LGCN

Regression coefficients: Y-variate = $\log_e$ CRR

	Estimate	SE	t
Constant	1.37	0.33	4.1
TEMP	0.10	0.01	13.1
LGCN	−0.40	0.10	−3.9

Analysis of variance

	Degrees of freedom	Sum of squares	Mean square
Regression	2	65.11	32.55
Residual	232	84.30	0.36
Total	234	149.41	0.64

Variation accounted for 43.1%

Table 6. Correlation matrices demonstrating the effect of sample size reduction on correlation coefficients

	ALAT	LONG	ALT	TEMP	RAIN	LOI	TOTN	EXTP	CA	EXTK	PH	SOILT	SOILM	EXTN	TOTP
Data sets = 231															
ALAT	1.00														
LONG	0.20	1.00													
ALT	−0.23	−0.41	1.00												
TEMP	−0.79	−0.48	−0.06	1.00											
RAIN	−0.81	−0.32	0.27	0.77	1.00										
LOI	−0.19	−0.02	−0.11	0.29	0.32	1.00									
TOTN	−0.09	−0.02	0.11	0.12	0.12	0.56	1.00								
EXTP	0.31	0.19	−0.20	−0.36	−0.45	−0.12	−0.07	1.00							
CA	0.08	−0.18	0.48	−0.25	−0.04	0.32	0.51	−0.07	1.00						
EXTK	−0.00	−0.02	0.20	−0.16	−0.06	0.19	0.05	0.66	0.17	1.00					
PH	0.16	0.31	−0.068	−0.40	−0.51	−0.49	−0.08	0.26	0.19	−0.03	1.00				
Data sets = 20															
ALAT	1.00														
LONG	1.00	1.00													
ALT	−0.32	−0.32	1.00												
TEMP	0.99	0.99	−0.38	1.00											
RAIN	−1.00	−1.00	0.32	−0.99	1.00										
LOI	−0.23	−0.23	−0.50	−0.12	0.23	1.00									
TOTN	0.18	0.18	−0.99	0.24	−0.18	−0.48	1.00								
EXTP	0.29	0.29	0.22	0.25	−0.29	0.17	−0.28	1.00							
CA	−0.04	−0.04	−0.72	0.03	0.04	−0.34	0.76	−0.72	1.00						
EXTK	0.24	0.24	0.09	0.23	−0.24	0.31	-0.13	0.87	-0.58	1.00					
PH	−0.17	−0.17	−0.87	−0.09	0.17	−0.31	0.93	−0.39	0.78	−0.21	1.00				
SOILT	−0.84	−0.84	−0.24	−0.80	0.84	−0.04	0.38	−0.42	0.45	−0.30	0.67	1.00			
SOILM	0.26	0.26	−0.99	0.30	−0.26	−0.60	0.98	−0.22	0.71	−0.11	0.87	0.30	1.00		
EXTN	0.65	0.65	−0.06	0.63	−0.65	−0.20	−0.03	0.21	−0.20	0.01	−0.27	−0.63	0.03	1.00	
TOTP	−1.00	−1.00	0.32	−0.99	1.00	0.23	−0.17	−0.29	0.04	−0.24	−0.18	0.84	−0.25	−0.65	1.00

adding a further 4 variables which are biologically important (period soil temperature, period soil moisture, extractable nitrogen, and total phosphorus) dramatically reduces the available data to only 20 sites. By referring to the correlation matrices given in Table 6, we can see that it would be difficult to interpret multiple regression analyses due to inter-correlations between the predictor variables. However, if we investigate pairs of predictor variables in more detail, a further difficulty becomes apparent.

In Figure 6 i, we plot absolute latitude against temperature, and most of the values appear in the top left of the Figure, with an indication that temperature decreases with latitude, as would be expected; however, the form of this relationship, whether it is assumed to be linear or not, will be highly affected by a set of 9 values at low latitude, with corresponding high temperature, appearing as a single cluster of points in the bottom right of the Figure. Even more dramatic discrepancies can be seen in Figures 6 ii and 6 iii, which plot absolute latitude against, respectively, altitude and loss-on-ignition. Both these Figures reveal a broad scatter of points, largely uncorrelated, at high latitudes, an absence of information at intermediate latitudes, and a confined spread of points at low latitudes; such distributions are far from a bivariate-normal distribution, and, in the cases of Figures 6 i and 6 iii, could give rise to apparent correlations between the 2 variables, which are largely dependent on the absence of intermediate values.

Similar discontinuous distributions can be seen for the biological parameters, as illustrated in Figures 7 i and 7 ii, which plot loss-on-ignition against, respectively, pH and total phosphorus. There is thus a dual problem with the data set: first, some of the biologically more interesting parameters are only recorded for a minority of sites, which may well not be representative (the percentage of variation of CRR explained for the subset cannot be expected to be the same for the whole data set); second, within the minority of sites for which some parameters are recorded, it appears that there are frequently sites which are atypical of the others; indeed, they may have been specifically selected as representing extreme, but interesting, circumstances; thus, any predictor variable for which these sites also have unusual values (as illustrated in Figures 6 & 7) could statistically act simply to distinguish these sites from the others (as binary variables would do), rather than indicating a true functional relationship between the parameter and CRR.

In Figure 8, we confirm that this possibility is a real risk with the present data set by showing scatter diagrams of CRR against 2 physical parameters, absolute latitude and temperature, and 2 chemical parameters, loss-on-ignition and pH. Examination of other scatter plots not shown in this paper shows that most of the regressor variables have very odd distributions with CRR, and that problems of outlying values are frequent. Indeed, the distributions of bivariate plots, and the frequency and magnitude of outlying sets are

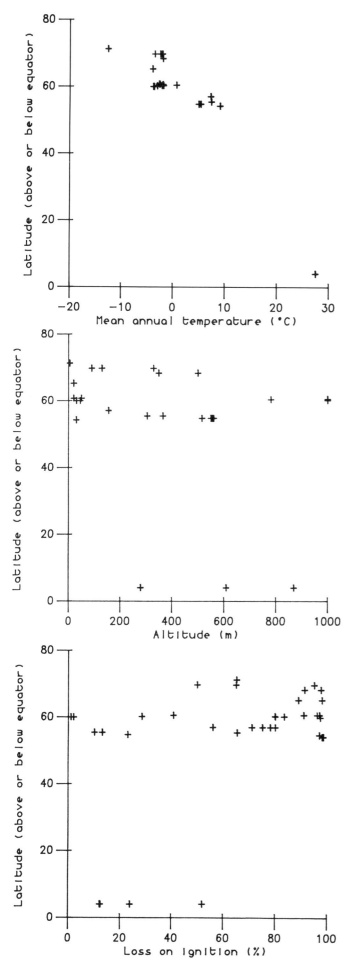

Figure 6. Scatter plots showing the relationship between absolute latitude and mean annual temperature, altitude, and loss-on-ignition

such that it is not possible to suggest useful non-linear transforms.

Unfortunately, the only feasible solution to these difficulties is to obtain more data to fill the gaps. If all the sites which were extreme in any parameter are omitted, the resulting data set is very small and almost certainly unrepresentative. Had the data set represented a more comprehensive series of sites (effectively filling in the gaps between the majority of sites and some of the extreme ones), a multivariate technique, such as principal component analysis, used on the predictor variables might perhaps have overcome statistical difficulties. However, principal component analysis does itself require a full data set in order to ordinate any one set, and would also be prone to giving undue weight to extreme values in any variable. Thus, in the analyses described below, we have felt obliged to confine our remarks to the broader trend exhibited by the fundamental physico-chemical parameters which are recorded for the great majority of data sets. Even so, our conclusions have to be modified in the light of which data sets were actually included in which analyses.

4.1.4 *Relationships with basic physico-chemical parameters*

During the workshop, a regression analysis on a large proportion of the data set indicated that 3 parameters, absolute latitude, temperature, and cotton strip depth, explained about 70% of the variance of CRR. During subsequent analyses, we noticed that this same equation explained 60% of the variation with a data set of 326, but only 34% with a data set reduced to 228 by the inclusion of additional interesting biological parameters. It appears, from a number of analyses, that temperature and absolute latitude are almost synonymous with regard to this data set, their inter-correlation explaining 60% of their variance. For several different, but large, subsets of data, absolute latitude and temperature are both highly significant predictors, but the accuracy of the prediction is not greatly affected by which is used, as might be expected given their high inter-correlation. For example, on a data set of 230, the relationship with the best 9 parameters (absolute latitude, temperature, rainfall, loss-on-ignition, total nitrogen, extractable phosphorus, calcium, total potassium and pH) is found to include temperature, rainfall and total nitrogen as having significant effects. However, the effects of rainfall and total nitrogen are only just significant ($P<0.05$, as opposed to the very highly significant effect of temperature, $P<0.001$), and the overall equation only explains 2% more variation than the correlation with temperature alone (which explains 35% variance).

A different, and somewhat larger, data set, with 325 degrees of freedom and using absolute latitude, temperature, rainfall and loss-on-ignition as predictor variables, ascribes highly significant effects to temperature, loss-on-ignition and rainfall, but not absolute

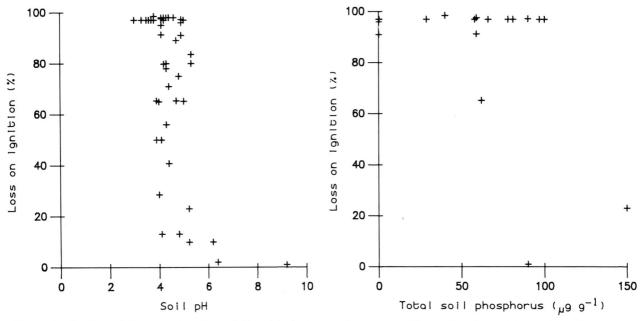

Figure 7. Scatter plots showing the relationship between loss-on-ignition with pH and total phosphorus

Table 7. Correlation matrix derived when attempting to provide a predictive equation for CRR, using 328 cases

	CRR	DEPTH	ALAT	TEMP	RAIN	LOI
CRR	1.00					
DEPTH	-0.20	1.00				
ALAT	-0.75	0.10	1.00			
TEMP	0.78	-0.14	-0.94	-1.00		
RAIN	0.48	-0.09	-0.72	0.79	1.00	
LOI	-0.65	0.07	0.50	-0.46	-0.19	1.00

Table 9. Correlation matrix derived when attempting to provide a predictive equation for CRR, using 235 cases

	CRR	DEPTH	ALAT	TEMP	RAIN	LOI	TOTN
CRR	1.00						
DEPTH	-0.30	1.00					
ALAT	-0.43	0.11	1.00				
TEMP	0.57	-0.19	-0.79	1.00			
RAIN	0.42	-0.06	-0.82	0.77	1.00		
LOI	-0.01	0.01	-0.13	0.23	0.20	1.00	
TOTN	0.07	-0.05	-0.06	0.09	0.06	0.57	1.00

Table 8. The analysis of variance table for the regression analysis shown in Table 7

Regression coefficients, Y-variate = CRR

	Estimate	SE	t
Constant	10.54	0.72	14.7
TEMP	0.55	0.04	14.2
LOI	-0.07	0.01	-9.5
RAIN	-0.00	0.00	-4.1
DEPTH	-0.11	0.04	-2.8

Analysis of variance

	Degrees of freedom	Sum of Squares	Mean Square
Regression	4	11939	2984.82
Residual	323	4332	13.41
Total	327	16271	49.76

Variation accounted for 73.0%

Table 10. The analysis of variance table for the regression analysis shown in Table 9

Regression coefficients, Y-variate = CRR

	Estimate	SE	t
Constant	2.42	0.26	9.4
TEMP	0.14	0.01	10.3
DEPTH	-0.05	0.01	-3.7
LOI	-0.01	0.00	-2.2

Analysis of variance

	Degrees of freedom	Sum of squares	Mean square
Regression	3	160.2	53.40
Residual	231	262.8	1.14
Total	234	423.0	1.81

Variation accounted for 37.1%

latitude, and explains 72% of the variation. Again, temperature explains a considerable proportion (60%) of the variation on its own. It is interesting to note that the larger data set (328 as opposed to 230) has nearly twice as much variance explained by the total model and by temperature alone (72% and 37% for the full model and 60% and 36% for temperature alone,

respectively). Temperature appears to be the most important of the physico-chemical parameters recorded, with loss-on-ignition and rainfall also having significant predictive effects, but the importance varying considerably with the sets of data that are included in a particular analysis.

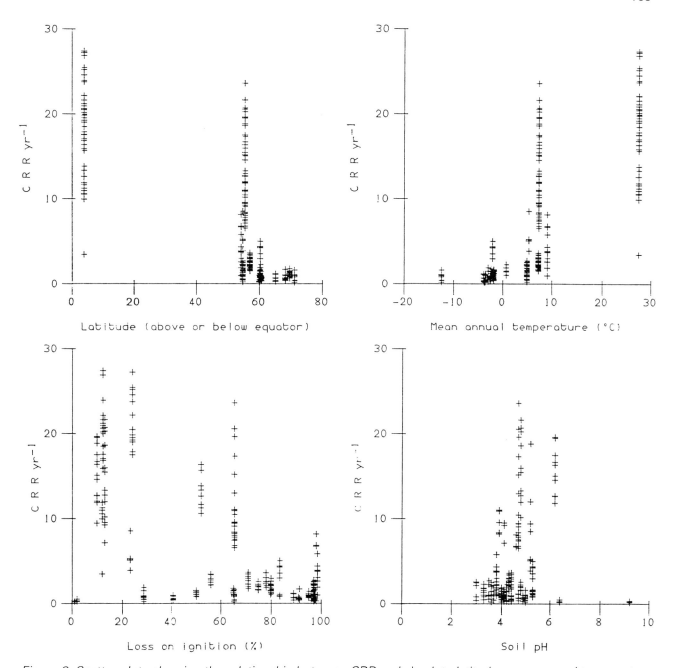

Figure 8. Scatter plots showing the relationship between CRR and absolute latitude, mean annual temperature, loss-on-ignition, and pH

In a final analysis, we attempted to find a good predictive equation for CRR based on variables which were recorded for 328 out of 329 of the data sets. CRR was predicted from depth of insertion, absolute latitude, temperature, rainfall, and loss-on-ignition. The first analysis included all 328 data sets; the correlation matrix is given in Table 7 and the analysis of variance for the regression analysis in Table 8. The 4 variables, temperature, loss-on-ignition, rainfall and depth, are shown to have significant effects, with slopes as given in Table 8. However, when the data set was reduced to two-thirds of this size, 235 data sets, and the analysis repeated, a somewhat different situation emerged. The correlation matrix is given in Table 9, and the analysis of variance for the final regression model in Table 10. The effect of rainfall is no longer significant, but the effects of temperature, depth and loss-on-ignition still are significant. However, it should be

noted that the parameter estimates are sensitive to the data sets used, although the coefficients for the effects of loss-on-ignition are not dissimilar between the 2 analyses (328 versus 235 sets). The parameter estimates for the effects of temperature and depth of insertion vary considerably, and significantly, between the 2 analyses.

We concluded, with regret, that the data set here analysed, although extensive, is not sufficiently comprehensive (lacking values for many important biological parameters for many of the sites investigated) to permit firm conclusions to be reliably drawn (over and above the broad trends evident from Tables 8 & 10). Accordingly, we have attempted to estimate some of the missing data using a compendium of meteorological data, which allows us to estimate, albeit somewhat imprecisely, some of the missing values on the basis

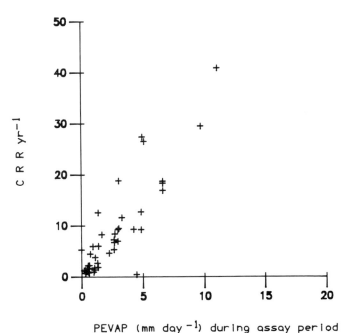

Figure 9. Scatter plot showing the relationship between CRR and period potential evaporation (PEVAP), using the augmented data set

Table 12. The analysis of variance table for the regression analysis of the extended data set

Regression coefficients, Y-variate = CRR

	Estimate	SE	T
Constant	0.20	0.92	0.2
PEVAP	3.13	0.26	12.2

Analysis of variance

	Degrees of freedom	Sum of Squares	Mean square
Regression	1	2796.0	2796.04
Residual	43	811.6	18.87
Total	44	3607.7	81.99

Variation accounted for 77.0%

Addition of other variables failed to improve significantly the explanation of residual variance, and Figure 9 shows the plot of CRR against PEVAP. Examination of the residuals failed to suggest any need for transformation, and they appeared randomly distributed.

5 Conclusions

One of the principal conclusions arising from this analysis of available cotton strip data is that it is very difficult to perform *a posteriori* multivariate analyses using a data set which is heterogeneous. The first part of the analysis revealed that apparent trends in the data were really a consequence of the few locations in which research effort had been made. This limitation severely affected the nature of the results, and their interpretation. It proved far more satisfactory to select a limited number of cases, examining these in more detail and extracting back-up information from published sources.

It is apparent from the analysis of the extended data set that the decomposition of cotton strips is strongly influenced by climate, and that the most useful climatic parameter for predicting cotton strip decay rate is PEVAP. In fact, the equation linking decomposition rate to PEVAP could be of use in deciding the time periods for inserting cotton strips at new sites.

The close relationship between PEVAP and decomposition rate is due to the fact that the 2 principal environmental controls on decomposition are temperature

of their geographic locations. This analysis is described below.

4.2 Augmented data set

Analysis of the original data set had highlighted the over-riding importance of climatic factors in determining the rate of decomposition of cotton strips. In the analyses of the extended data set, soil chemical variables were not included, and period temperatures, rainfall and potential evaporation (PEVAP) were derived from published tables.

Table 11 shows the correlation matrix between the variables used in the multiple regression analysis, in which CRR was the dependent variable. From this matrix, the importance of PEVAP in explaining decomposition rate is apparent, and this importance is reinforced in Table 12, which shows the analysis of variance table and regression coefficients resulting from the step-wise multiple regression. Period potential evaporation explained 77% of the variance, yielding the equation:

$$CRR = 0.20 + 3.13 \text{ (PEVAP)}$$

Table 11. Correlation matrix derived when attempting to derive a predictive equation for CRR, using the restricted data set (df = 43)

	TS	CRR	DAYS	LAT	LONG	TEMP	PTEMP	RAIN	PRAIN	PEVAP	ALT
TS	1.00										
CRR	−0.16	1.00									
DAYS	−0.11	−0.68	1.00								
LAT	−0.24	−0.67	0.62	1.00							
LONG	−0.11	−0.07	0.11	−0.17	1.00						
TEMP	0.22	0.67	−0.70	−0.94	0.29	1.00					
PTEMP	−0.07	0.69	−0.70	−0.78	0.28	0.83	1.00				
RAIN	0.02	0.62	−0.67	−0.83	0.30	0.85	0.72	1.00			
PRAIN	−0.05	0.67	−0.63	−0.60	0.21	0.71	0.70	0.76	1.00		
PEVAP	0.00	0.88	−0.63	−0.78	−0.07	0.77	0.80	0.62	0.67	1.00	
ALT	−0.16	−0.39	0.45	0.37	0.15	−0.51	−0.38	−0.32	−0.40	−0.47	1.00

and moisture (Bunnell *et al.* 1977). The magnitude of PEVAP is dictated by rainfall and temperature, being large when rainfall and temperature are high, and lower when either of these factors decrease. Thus, conditions of high PEVAP are those which favour a high decomposition rate.

The current synthesis of data suggests that decomposition of a single substrate, when placed in a widely diverging range of environments, can be modelled with accuracy from readily available climatic data. Swift *et al.* (1979) suggested that decomposition rate is the product of the interaction of substrate quality, physical environment and decomposer organisms. Hill *et al.* (1985) concluded, from a mathematical model of the decay process of cotton in soil, that the rate of degradation depended mainly on the physico-chemical environment in the soil, and was insensitive to the size of the microbial inoculum.

In the synthesis presented here, the substrate quality has been kept constant between sites, and differences in cellulose decomposition rates at different sites are, therefore, due to a combination of physico-chemical environment and presence of decomposer organisms. The analysis further suggests that, of these 2 factors, a major component determining decay rate is physical environment, particularly temperature and moisture, as reflected by PEVAP. Residual values may well be a combination of discrepancies between published PEVAP values and actual site values, differences in site chemistry, or a measure of decomposer population differences. The simplest hypothesis necessary to explain the observed results is to assume that wherever a cotton strip is placed, it will select for a cellulolytic flora capable of cotton degradation, and that the subsequent rate of decay becomes a simple function of physico-chemical environment.

A major feature lacking in the current analysis is the detailed description of soil chemistry at each of the sites and, if the cotton strip method is to be used to increase our understanding of the controlling factors for cotton decay around the world, it is essential that workers make a full assessment and record of climatic and chemical parameters at their sites.

The results reported here lend support to the observations of Meentemeyer (1978) that decomposition on a regional scale relates to evaporative losses. The parameter investigated by Meentemeyer (1978) was actual evapotranspiration, to which potential evaporation, as used in this study, approximates. The more recent work of Berg *et al.* (1984) confirms that this approach is useful for furthering our understanding of the decomposition of standard litter material from Scots pine (*Pinus sylvestris*), yet he also emphasizes that local site variability, acting both through litter quality and microscale climate, can be extremely important in determining litter decomposition rates (Berg *et al.* 1982).

We emphasize that the rate of litter decay is determined by the *interaction* of substrate quality, physico-chemical environment and decomposer population. Although the use of a standard substrate can provide insights into the factors controlling the rates of decomposition, it is currently impossible to extrapolate such results to the natural substrates intrinsic to the site.

6 Acknowledgements
Without the ready co-operation of the following people in supplying cotton strip data and information on the sites (Figure 1), this analysis could not have been done: D D French (Glendye), G Howson (Elwy), P M Latter (Kershope), A Ligget (Branden), E Maltby (Everglades, Pearl River, Red River, Brazil), J Proctor (Sabah, Malaysia), K Van Cleve (Alaska), P J Vickery (Sourhope), Professor Y Waisel (Tel Aviv), D D Wynn-Williams (Faraday, Signy).

7 References
Abrahamsen, G., Hornvedt, R. & Tveite, B. 1975. *Impacts of acid precipitation on coniferous forest ecosystems.* (SNSF-project research report FR 2/75.) Aas: SNSF Project, Norwegian Forest Research Institute.
Alvey, N.G., Banfield, C.F., Baxter, R.I. *et al.* 1983. *Genstat manual.* Harpenden: Rothamsted Experimental Station.
Berg, B., Hannus, K., Popoff, T. & Theander, O. 1982. Changes in organic chemical components of needle litter during decomposition: long-term decomposition in a Scots pine forest. 1. *Can. J. Bot.,* **60,** 1310-1319.
Berg, B., Jansson, P. & Meentemeyer, V. 1984. Litter decomposition and climate — regional and local models. In: *State and change of forest ecosystems — indicators in current research,* edited by G.I. Agren, 369-404. Uppsala: Swedish University of Agricultural Sciences.
Bunnell, F.L., Tait, D.E.N., Flanagan, P.W. & Van Cleve, K. 1977. Microbial respiration and substrate weight loss. 1. A general model of the influences of abiotic variables. *Soil Biol. Biochem.,* **9,** 33-40.
Esser, G., Asselmann, I. & Lieth, H. 1982. Modelling the carbon reservoir in the system compartment 'litter'. *SCOPE/UNEP Sonderband,* **52,** 39-58.
Heal, O.W., Howson, G., French, D.D. & Jeffers, J.N.R. 1974. Decomposition of cotton strips in tundra. In: *Soil organisms and decomposition in tundra,* edited by A.J. Holding, O.W. Heal, S.F. MacLean & P.W. Flanagan, 341-362. Stockholm: Tundra Biome Steering Committee.
Hill, M.O., Latter, P.M. & Bancroft, G. 1985. A standard curve for inter-site comparisons of cellulose degradation using the cotton strip method. *Can. J. Soil Sci.,* **65,** 605-619.
Hill, M.O., Latter, P.M. & Bancroft, G. 1988. Standardization of rotting rates by a linearizing transformation. In: *Cotton strip assay: an index of decomposition in soils,* edited by A.F. Harrison, P.M. Latter & D.W.H. Walton, 21-24. (ITE symposium no. 24.) Grange-over-Sands: Institute of Terrestrial Ecology.
Howson, G. 1988. Use of the cotton strip assay to detect potential differences in soil organic matter decomposition in forests subjected to thinning. In: *Cotton strip assay: an index of decomposition in soils,* edited by A.F. Harrison, P.M. Latter & D.W.H. Walton, 94-98. (ITE symposium no. 24.) Grange-over-Sands: Institute of Terrestrial Ecology.
Meentemeyer, V. 1978. Macroclimate and lignin control of litter decomposition rates. *Ecology,* **59,** 465-472.
Müller, M.J. 1982. *Selected climatic data for a global set of standard stations for vegetation science.* The Hague: Junk.
Rosswall, T. 1974. Cellulose decomposition studies on the tundra. In: *Soil organisms and decomposition in tundra,* edited by A.J. Holding, O.W. Heal, S.F. MacLean & P.W. Flanagan, 325-340. Stockholm: Tundra Biome Steering Committee.
Swift, M.J., Heal, O.W. & Anderson, J.M. 1979. *Decomposition in terrestrial ecosystems.* Oxford: Blackwell Scientific.

Appendix I
Current method for preparation, insertion and processing of cotton strips

Plate 14 illustrates diagrammatically the use of the assay.

1 Material
Shirley Soil Burial Test Fabric should normally be used (Walton & Allsopp 1977; Sagar 1988). This is 100% combed cotton, with coloured marker threads in the warp direction to aid cutting and fraying of samples to size (Plate 7).

2 Preparation
For field studies, the usual size of strip is 35 cm (weft) x 10 cm (warp). The strips are torn, or cut, from the cloth. Initial washing of the Shirley cloth is not necessary as contamination is kept to a minimum during production. Other sizes of piece are used according to the purpose and method of the experiment (Collins et al. 1988; Holter 1988).

Packs of 5 strips are wrapped together in aluminium foil, and sterilized in an autoclave at 121° for 15 min, to avoid the introduction of organisms alien to a soil. Some 3–4 kg gain in tensile strength occurs, possibly due to relaxation of strains put into the fabric during the finishing processes, or to alteration to the thread crimp of the cloth (B F Sagar pers. comm.), but the variability in tensile strength (TS) is not increased (Latter et al. 1988).

3 Field placing
For investigation down a soil profile, strips are inserted vertically using a spade (Plate 6). A sharp knife is used to check for any roots or stones which may impede subsequent insertion and to cut through the litter layer to diminish carry-down of surface material during insertion. Then, folding the lower 2–3 cm of strip over the blade of a small straight spade, the strip is pushed vertically into the soil, leaving approximately 4 cm protruding above the surface. In most soils, it is necessary to ease removal of the spade by sliding it sideways a few times, while the strip should remain in place. The creation of a gap round the strip, which can be caused by moving the spade forward or backwards, is to be avoided. The soil is then pressed back against the strip after insertion, using the feet or hands according to the nature of the surface.

To reduce mechanical stretching of cloth or potential damage from stones, some workers first make a slot in the soil with a sharpened steel plate and then insert the strip into this slot using a blunt-edged former (Walton 1988). Maltby (1988) has also devised a special tool to aid insertion in bogs. The aim in all cases must be to minimize disturbance of the site and to attain close contact between strip and soil. For sites with a gritty or stony subsoil near the surface, French and Howson (1982) placed strips round the base of a block of the surface peat, calling this the horizon method. Lawson (1988), on the other hand, placed the 2 ends of strip into the soil.

Where interference by animals is expected, the protruding top of the strip can be firmly pinned down or the strip completely inserted (or trimmed down), and the position marked in some way; alternatively, a cage of plastic-coated wire can be firmly fixed over the position (Vickery & Floate 1988). Plastic rods, 30 cm long, are frequently used to mark positions; these are inert and bend without breaking.

4 Controls
In the early stages of experiments with the unbleached calico, it became apparent that the variability of the tensile strength of undecayed cloth was considerable and that systematic controls at all stages of use were vital. They were needed to assess the effects of batch of cloth, autoclaving, washing, the mechanical stretching which occurred during field insertion, and of soil impregnation or cementation. The controls allowed assessment of the minimum detectable change in TS that could be attributed to cellulose decay in the soil.

Cloth controls are unburied samples used to check batch variation and effect of sterilization or storage of cloth. The TS of fabric controls was not found to vary in any pattern throughout the width and length of a single roll of cloth. However, Howson (1988) has found variation between batches of the earlier Shirley cloth.

Field controls (also called insertion controls) are most important because they provide a baseline for expression of TS loss of cotton (CTSL) and allow for errors due to insertion and retrieval and for any TS increase due to cementation in particular soils (French 1988). Field controls are inserted and removed on the same day, usually at the time of retrieval, but, in some extreme conditions where soil may be in a different condition at insertion and retrieval dates, eg frozen or very dry, then 2 sets of field controls may be needed to cover the range of conditions at both times. They should be placed on each plot type, or soil, at each sampling. Field controls are then washed and tested along with the test samples.

5 Retrieval time
Cotton strips are left in the field until the tensile

strength has reached approximately 50% loss, but does not exceed 85%. A reasonable estimate of assay period for any site can be obtained by using the climatic compilation of Müller (1982) to derive climatic data for the nearest comparable site, and then applying the equation relating cotton rotting rate (CRR) to potential evaporation (PEVAP) given in Ineson et al. (1988), then

$$\text{days to 50\% CTSL} = \frac{365}{\text{CRR}}$$

Alternatively, a set of time controls should be inserted in representative parts of a site, and, if considerable variation is to be expected, the suspected most active and inactive areas would be included. A few strips are retrieved and TS is tested at intervals to estimate retrieval time using the linearizing formula (Hill et al. 1985, 1988).

6 Retrieval method

The soil surface level is marked with a permanent marker pen (eg Pentel N50) or wax pencil. The retrieval method will vary according to the nature of the site, but should aim to avoid pulling the strip, particularly in the 1–10 cm region which is likely to be the most decomposed.

A method found effective in forest sites uses 2 small 'ladies' garden forks, which are used as in separating roots of perennial plants. The litter layer is first pulled back and the 2 forks inserted with tines 4 cm from the strip on both sides. Holding the shafts vertical, the forks are pushed into the soil and, by crossing the handles and pulling apart, a gap is achieved round the strip. The base of the strip can then be reached, and the strip lifted out by hand. Soil and litter are pressed back into position after retrieval, causing minimal disturbance.

In the original method (Latter & Howson 1977), a cut is made away from the sides of the strip with a sharp knife and then at right angles to a distance of 3–4 inches from one face of strip. The soil can then be removed from this face using a spade or trowel, so the base of the strip is reached and removed as before. On some sites, a fork or spade used at right angles to the side of the strip is effective.

Strips are labelled using a permanent marker pen or paper tags stapled to the top of the strip, or base if there is no free top. Strips are placed between sheets of blotting paper and wrapped in foil packets for return to the laboratory. Any storage at this stage should be at 0–2°C for a minimum time, not exceeding one week.

The strips are washed to remove soil in order to reduce CTSL during storage and maintain cleanliness of tensometer and jaws. A jet of water is sufficient, and a washing box has been made at the Institute of Terrestrial Ecology's Merlewood Research Station for this purpose and for washing roots (Benham 1984). No scrubbing treatment should be used as this causes TS changes (Latter et al. 1988). At this stage, the British Standard 6085 (Anon 1981) recommends soaking in 70% ethanol at room temperature for 4 h to minimize health hazards. Strips are dried at room temperature and any storage is under dry conditions. Dried strips showed CTSL of less than 0.4 kg per week.

7 Preparation for tensile testing

The procedures for tensile testing follow the British Standard 2576 (Anon 1986) as far as possible. To provide values for different levels of a profile, strips are cut horizontally at intervals (4 cm or 2 cm are often used), labelled with suitable felt-tip pen (Berol notewriter) and frayed down 0.5 cm on each side. Use of narrower pieces increased the error due to isolated points of high decomposition, but they can also show a lower mean CTSL. For examination of closer depth intervals, a sequence of strips could be placed horizontally. Although recommended in the British Standard 2576 (Anon 1986), to prevent escape of longtitudinal threads during testing, fraying is not absolutely essential for the field test (Latter et al. 1988). Because fraying of large numbers can create a health hazard in unventilated areas, strips can be prepared by cutting accurately to near the size required and removing only one or 2 threads from each side to obtain an exact size and even edge with undamaged threads. If the size corresponds with the coloured marker threads, one coloured thread is removed from each side to obtain an accurate number of weft threads to each test piece. Visual observation has indicated that decomposition may be inhibited in some soils by the dyes used for the marker threads, particularly the brown, as these threads were sometimes left intact among the more decomposed undyed parts of retrieved samples. The green dye also became solubilized in some soils, probably due to microbial action. The current Shirley cloth has blue marker threads. No further work has been done on these aspects.

Plate 13. Machine used for fraying substrips (Photograph G Howson)

168

INSERT

Sterilized strips

Cut through litter layer

Insert in soil

Remove spade

RETRIEVE

Mark soil level

Label top
of strip

Remove with forks or spade.
Repack strips in foil

WASH

Wash with jet
of water, by
hand or in
washing
apparatus

Dry in air, repack in foil   STORE DRY STRIPS

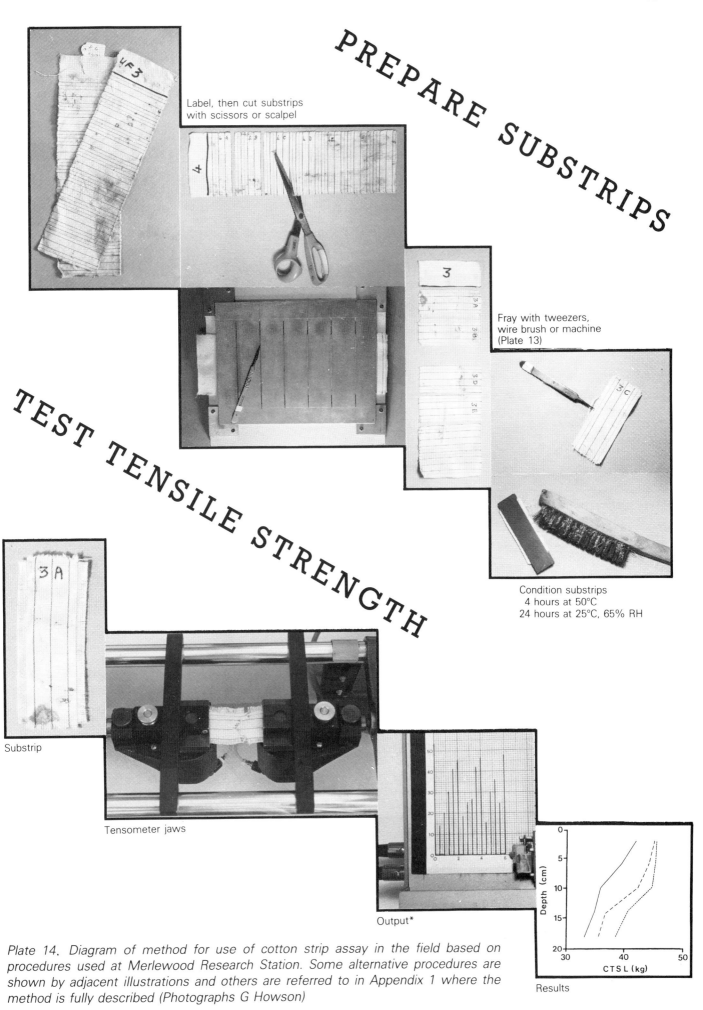

PREPARE SUBSTRIPS

Label, then cut substrips
with scissors or scalpel

Fray with tweezers,
wire brush or machine
(Plate 13)

TEST TENSILE STRENGTH

Condition substrips
4 hours at 50°C
24 hours at 25°C, 65% RH

Substrip

Tensometer jaws

Output*

Results

Plate 14. Diagram of method for use of cotton strip assay in the field based on procedures used at Merlewood Research Station. Some alternative procedures are shown by adjacent illustrations and others are referred to in Appendix 1 where the method is fully described (Photographs G Howson)

* The chart recorder output is modified to compress the extension curve, normally shown, to a single vertical line indicating only the breaking load. This modification allows the output of many successive results on a single chart

Fraying can be done with fine tweezers, a wire brush, or with a purpose-built machine made at Merlewood (Plate 13), and can then be a quick and accurate way of preparing strips.

Prior to tensile testing, strips are conditioned by drying at 50°C for 4 hours, followed by 24 hours in a humidity oven at 65% relative humidity, 20°C, immediately before testing, as described in British Standard 6085 (Anon 1981). Several workers (Walton & Allsopp 1977; Wynn-Williams 1988) test wetted strips to avoid the need for humidity control.

8 Recording of decomposition

Tensile strength loss is used as the measure of decomposition (Sagar 1988) and is measured in the warp direction on a tensometer. This can be a simple weight apparatus (Mills et al. 1972), or constructed on a balance and sliding weight principle (P S Rhodes pers. comm.), or a commercial tensometer.

A primary problem of tensile testing is the use of suitable tensometer jaws to grip the test material. For cotton and for the small size of piece (in comparison to the standard textile test) used in this method, some modification of jaws may be necessary, and leather (chamois) or rubber linings have been found effective for mechanical jaws. Walton (1985) showed no difference in results using different mechanical jaw types. The 5 cm wide rubber-lined jaws on the Monsanto pneumatic unit are very quick and easy to use, and enable us to test up to 300 samples in a day. The air pressure supply for the pneumatic jaws is maintained at 60 lb in^{-2}. Provided pressure was not below 50 lb in^{-2}, there was no significant TS change with increase to 70 lb in^{-2}.

The CTSL can be expressed as a rate by a cube root transformation (Hill et al. 1988), with results given as time (in days or weeks) to 50% CTSL or as number of strips decomposing to 50% CTSL per period of time, usually one year (Figure 1).

Under some conditions, an increase of tensile strength can occur and may be due to soil impregnation or the growth of certain fungi (French 1988; Latter et al. 1988).

Experimental studies have shown some relationship with the weight loss of strips (Latter & Walton 1988) when both are expressed as time to 50% loss. Weight is lost at a slower rate and TS loss reaches nearly

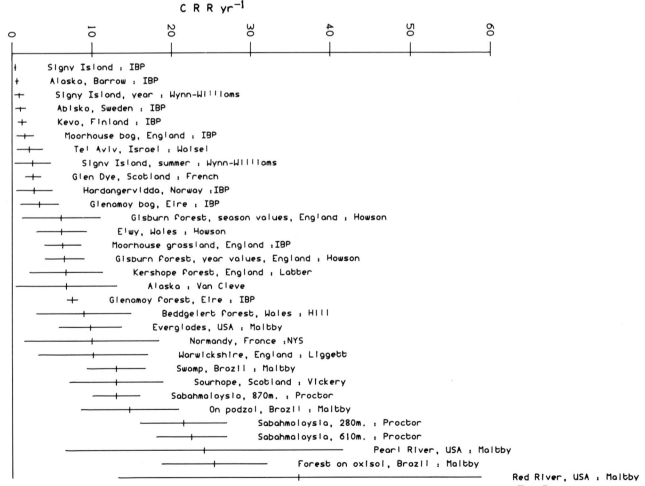

Figure 1. Ranges of reported CRR values for world-wide sites, also showing the high variability on the more active sites. It should be noted that only a few values represent a full year's data, most values being influenced by the seasonal climate of the respective insertion periods. The majority of sites are indicated in Figure 1 of Ineson et al. (1988), and the full names and addresses of the respective workers (given after site names) appear in Appendix IV

100%, while weight loss is approximately 25%. The 2 parameters are, to some extent, based on different aspects of cellulose decomposition (Howard 1988; Sagar 1988), and a close relationship is not to be expected.

When carried out according to these instructions, many ecologists have found the assay a useful addition to their studies, giving information on comparative decomposition activity in ecological investigations. The assay is simple and easy to use on any site with only slight modifications of insertion/retrieval method to suit site conditions. Any careful worker can quickly be trained.

The current cost of cloth from the Shirley Institute is £6 per metre, and the approximate time in minutes taken to process one cotton strip of 5 profile subsamples (excluding travelling time) is as follows:

Tear, pack and sterilize	1.5
Place in field	2.5
Retrieve from field	2.8
Wash, dry and re-pack	2.4
Cut, label and fray	10.0
Tensile test and write out data	6.0

The availability of a tensometer is likely to be the main problem, but these machines are available in many universities. At the present time, the Monsanto machine at Merlewood can be used for a bench fee.

9 References

Anon. 1981. *Method of test for the determination of the resistance of textiles to microbiological deterioration.* (BS 6085.) London: British Standards Institution.

Anon. 1986. *Methods of test for textiles — woven fabrics — and elongation (strip method).* (BS 2576.) London: British Standards Institution.

Benham, D.G. 1984. Vegetation/root washing machine. *Annu. Rep. Inst. terr. Ecol. 1983,* 88-89.

Collins, H., Gitay, H., Scandrett, E. & Pearce, T. 1988. Decomposition rates in the ericaceous belt of Mount Aberdare, Kenya. In: *Cotton strip assay: an index of decomposition in soils,* edited by A.F. Harrison, P.M. Latter & D.W.H. Walton, 123-125. (ITE symposium no. 24.) Grange-over-Sands: Institute of Terrestrial Ecology.

French, D.D. 1988. The problem of cementation. In: *Cotton strip assay: an index of decomposition in soils,* edited by A.F. Harrison, P.M. Latter & D.W.H. Walton, 32-33. (ITE symposium no. 24.) Grange-over-Sands: Institute of Terrestrial Ecology.

French, D.D. & Howson, G. 1982. Cellulose decay rates measured by a modified cotton strip method. *Soil Biol. Biochem.,* **14,** 311-312.

Hill, M.O., Latter, P.M. & Bancroft, G. 1985. A standard curve for inter-site comparison of cellulose degradation using the cotton strip method. *Can. J. Soil Sci.,* **65,** 609-619.

Hill, M.O., Latter, P.M. & Bancroft, G. 1988. Standardization of rotting rates by a linearizing transformation. In: *Cotton strip assay: an index of decomposition in soils,* edited by A.F. Harrison, P.M. Latter & D.W.H. Walton, 21-24. (ITE symposium no. 24.) Grange-over-Sands: Institute of Terrestrial Ecology.

Holter, P. 1988. Cellulolytic activity in dung pats in relation to their disappearance rate and earthworm biomass. In: *Cotton strip assay: an index of decomposition in soils,* edited by A.F. Harrison, P.M. Latter & D.W.H. Walton, 72-77. (ITE symposium no. 24.) Grange-over-Sands: Institute of Terrestrial Ecology.

Howard, P.J.A. 1988. A critical evaluation of the cotton strip assay. In: *Cotton strip assay: an index of decomposition in soils,* edited by A.F. Harrison, P.M. Latter & D.W.H. Walton, 34-42. (ITE symposium no. 24.) Grange-over-Sands: Institute of Terrestrial Ecology.

Howson, G. 1988. Use of the cotton strip assay to detect potential differences in soil organic matter decomposition in forests subjected to thinning. In: *Cotton strip assay: an index of decomposition in soils,* edited by A.F. Harrison, P.M. Latter & D.W.H. Walton, 94-98. (ITE symposium no. 24.) Grange-over-Sands: Institute of Terrestrial Ecology.

Ineson, P., Bacon, P.J. & Lindley, D.K. 1988. Decomposition of cotton strips in soil: analysis of the world data set. In: *Cotton strip assay: an index of decomposition in soils,* edited by A.F. Harrison, P.M. Latter & D.W.H. Walton, 155-165. (ITE symposium no. 24.) Grange-over-Sands: Institute of Terrestrial Ecology.

Latter, P.M. & Howson, G. 1977. The use of cotton strips to indicate cellulose decomposition in the field. *Pedobiologia,* **17,** 145-155.

Latter, P.M. & Walton, D.W.H. 1988. The cotton strip assay for cellulose decomposition studies in soil: history of the assay and development. In: *Cotton strip assay: an index of decomposition in soils,* edited by A.F. Harrison, P.M. Latter & D.W.H. Walton, 7-10. (ITE symposium no. 24.) Grange-over-Sands: Institute of Terrestrial Ecology.

Latter, P.M., Bancroft, G. & Gillespie, J. 1988. Technical aspects of the cotton strip assay method. *Int. Biodeterior,* **24.** In press.

Lawson, G.J. 1988. Using the cotton strip assay to assess organic matter decomposition patterns in the mires of South Georgia. In: *Cotton strip assay: an index of decomposition in soils,* edited by A.F. Harrison, P.M. Latter & D.W.H. Walton, 134-138. (ITE symposium no. 24.) Grange-over-Sands: Institute of Terrestrial Ecology.

Maltby, E. 1988. Use of cotton strip assay in wetland and upland environments — an international perspective. In: *Cotton strip assay: an index of decomposition in soils,* edited by A.F. Harrison, P.M. Latter & D.W.H. Walton, 140-154. (ITE symposium no. 24.) Grange-over-Sands: Institute of Terrestrial Ecology.

Mills, J., Allsopp, D. & Eggins, H.O.W. 1972. Some new developments in cellulosic material testing using perfusion techniques. In: *Biodeterioration of materials Vol II,* edited by A.H. Walters & E.H.H. Van Der Plas, 227-232. Barking, Essex: Applied Science.

Müller, M.J. 1982. *Selected climatic data for a global set of standard stations for vegetation science.* The Hague: Junk.

Sagar, B.F. 1988. The Shirley Soil Burial Test Fabric and tensile testing as a measure of biological breakdown of textiles. In: *Cotton strip assay: an index of decomposition in soils,* edited by A.F. Harrison, P.M. Latter & D.W.H. Walton, 11-16. (ITE symposium no. 24.) Grange-over-Sands: Institute of Terrestrial Ecology.

Vickery, P.J. & Floate, M.J.S. 1988. Effects of lime and pasture improvement on cotton strip decomposition in 3 Scottish acid hill soils. In: *Cotton strip assay: an index of decomposition in soils,* edited by A.F. Harrison, P.M. Latter & D.W.H. Walton, 109-112. (ITE symposium no. 24.) Grange-over-Sands: Institute of Terrestrial Ecology.

Walton, D.W.H. 1985. Tensometer and jaw types in testing textiles for tensile strengths. *Int. Biodeterior.,* **21,** 301-302.

Walton, D.W.H. 1988. Problems and advantages of using the cotton strip asay in polar and tundra sites. In: *Cotton strip assay: an index of decomposition in soils,* edited by A.F. Harrison, P.M. Latter & D.W.H. Walton, 43-45. (ITE symposium no. 24.) Grange-over-Sands: Institute of Terrestrial Ecology.

Walton, D.W.H. & Allsopp, D. 1977. A new test cloth for soil burial trials and other studies on cellulose decomposition. *Int. Biodeterior. Bull.,* **13,** 112-115.

Wynn-Williams, D.D. 1988. Cotton strip decomposition in relation to environmental factors in the maritime Antarctic. In: *Cotton strip assay: an index of decomposition in soils,* edited by A.F. Harrison, P.M. Latter & D.W.H. Walton, 126-133. (ITE symposium no. 24.) Grange-over-Sands: Institute of Terrestrial Ecology.

Appendix II
Use of cotton in ecological studies
List of reference papers prior to publication of Symposium proceedings

Agarwal, C.P. & Chauhan, R.K.S. 1976. *Neurospora tetrasperma*: its occurrence in Indian soil and its cellulolytic activity. *Proc. Indian nat. Sci. Acad. B*, **42**, 122-124.

Baker, J.H. 1974. Comparison of the microbiology of four soils in Finnish Lapland. *Oikos*, **25**, 209-215.

*****Beiderbeck, V.O., Paul, E.A., Lowe, W.E., Shields, J.A. & Willard, J.R.** 1974. *Soil microorganisms: II. Decomposition of cellulose and plant residues.* (Matador Project technical report no. 39.) Saskatoon: University of Saskatchewan.

Brown, A.H.F., Gardener, C.L. & Howson, G. 1973. Differences in some biological attributes of soils developed under four tree species. In: *Biological processes and soil fertility*, edited by J. Tinsley & D.S. Jenkinson, 58. Reading: British Society of Soil Science.

Davis, R.C. 1986. Environmental factors influencing decomposition rates in two antarctic moss communities. *Polar Biol.*, **5**, 95-103.

Dighton, J., Thomas, E.D. & Latter, P.M. 1987. Interactions between tree roots, mycorrhizas, a saprotrophic fungus and the decomposition of organic substrates in a microcosm. *Biol. Fertil. Soils*, **4**, 145-150.

French, D.D. & Howson, G. 1982. Cellulose decay rates measured by a modified cotton strip method. *Soil Biol. Biochem.*, **14**, 311-312.

French, D.D. 1984. The problem of 'cementation' when using cotton strips as a measure of cellulose decay in soils. *Int. Biodeterior.*, **20**, 169-172.

Greaves, M.P., Cooper, S.L., Davies, H.A., Marsh, J.A.P. & Wingfield, G.I. 1978. *Methods of analysis for determining the effects of herbicides on soil micro-organisms and their activities.* (Technical report no. 45.) Yarnton: Weed Research Organization.

Grossbard, E. 1973. Rapid techniques for the assessment of the effect of herbicides on soil microorganisms and cellulolytic activity. In: *Modern methods of the study of microbial ecology*, edited by T. Rosswall, 473-474. Stockholm: Swedish Natural Science Research Council.

Grossbard, E. 1975. Techniques for the assay of the effects of herbicides on the soil microflora. In: *Some methods for microbial assay*, edited by R.G. Board & D.W. Lovelock, 235-256. London: Academic Press.

Grossbard, E. & Wingfield, G.I. 1975. The effect of herbicides on cellulose decomposition. In: *Some methods for microbial assay*, edited by R.G. Board & D.W. Lovelock, 235-236. London: Academic Press.

Heal, O.W., Flanagan, P.W., French, D.D. & MacLean, S.F. Jr. 1981. Decomposition and accumulation of organic matter. In: *Tundra ecosystems: a comparative analysis*, edited by L.C. Bliss, O.W. Heal & J.J. Moore, 587-634. Cambridge: Cambridge University Press.

Heal, O.W., Howson, G., French, D.D. & Jeffers, J.N.R. 1974. Decomposition of cotton strips in tundra. In: *Soil organisms and decomposition in tundra*, edited by A.J. Holding, O.W. Heal, S.F. MacLean & P.W. Flanagan, 341-362. Stockholm: Tundra Biome Steering Committee.

Heal, O.W., Latter, P.M. & Howson, G. 1978. A study of the rates of decomposition of organic matter. In: *Production ecology of British moors and montane grasslands*, edited by O.W. Heal & D.F. Perkins, 136-159. Berlin: Springer.

Hornung, M. 1986. Soil data collection and interpretation in some current projects in the Institute of Terrestrial Ecology. *Rep. Welsh Soils Discuss. Grp*, no. 22, 67-98.

Hill, M.O., Latter, P.M. & Howson, G. 1985. A standard curve for inter-site comparison of cellulose degradation using the cotton strip method. *Can. J. Soil Sci.*, **65**, 609-619.

Latter, P.M. & Howson, G. 1977. The use of cotton strips to indicate cellulose decomposition in the field. *Pedobiologia*, **17**, 145-155.

Liggett, A.C. 1983. *Comparative assessment of the cellulase activity in the soils of Brandon Marsh Nature Reserve.* BSc thesis, Coventry (Lanchester) Polytechnic.

Maltby, E. 1985. Effects of nutrient loadings on decomposition profiles in the water column and submerged peat in the Everglades. In: *Tropical peat resources — prospects and potential. Proc. int. Peat Soc. Symp.*, 450-464.

Maltby, E. 1986. Investigations of the peat cover of the North York Moors. In: *Moorland management*, edited by D.C. Statham, 28-41. Helmsley, Yorkshire: North York Moors National Park.

Maltby, E. 1986. *Soil property responses to surface treatment and associated management practices at Halscombe Allotment/Humbers Ball and Old Barrow Down/Hawkridge Plain 1978–1986.* Exmoor National Park Authority.

Miles, J. & Young, W.F. 1980. The effects on heathland and moorland soils in Scotland and northern England following colonization by birch (*Betula* spp.). *Bull. Ecol.*, **11**, 233-242.

Miles, J. 1981. *Effect of birch of moorlands.* Cambridge: Institute of Terrestrial Ecology.

Miles, J. 1984. *Gruinard Island. Assessment of ecological consequences of decontamination. Phase II.* (Natural Environment Research Council contract report to the Ministry of Defence.) Banchory: Institute of Terrestrial Ecology. (Unpublished.)

Miles, J., Latter, P.M., Smith, I.R. & Heal, O.W. 1988. Ecological effects of killing *Bacillus anthracis* on Gruinard Island with formaldehyde. *Reclam. Reveg. Res.* In press.

Mirzoev, O.G. 1973. Cellulose decomposition in the litter/humus of birch/*Carex pilosa* stands of different ages. *Lesovedenie*, no. 1, 86-88.

Rovira, A.D. 1953. A study of the decomposition of organic matter in red soils of the Lismore district. *Aust. Conf. Soil Sci., Adelaide*, **1**, 3.17, 1-4.

*****Smagina, M.V.** 1984. The activity of cellulose-decomposing micro-organisms in peat-bog soils. *Lesovedenie*, no. 1, 52-58.

Smith, M.J. & Walton, D.W.H. 1988. Patterns of cellulose decomposition in four subantarctic soils. *Polar Biol.* In press.

Springett, J.A. 1971. The effects of fire on litter decomposition and on the soil fauna in a *Pinus pinaster* plantation. In: *4th Colloquium Pedobiologiae, Dijon, 1970*, 529-535. Paris: Institut National de la Recherche Agronomique.

Springett, J.A. 1979. The effects of a single hot summer fire on soil fauna and on litter decomposition in jarrah (*Eucalyptus marginata*) forest in Western Australia. *Aust. J. Ecol.*, **4**, 279-291.

Springett, J.A. 1980. The use of cotton strips in a microcosm study of the energy cost of a predator–prey relationship. In: *Soil biology as related to land use practices*, edited by D.L. Dindal, 637-642. Washington, DC: Office of Pesticide and Toxic Substances, EPA.

Tsutsumi, T. & Katagiri, S. 1974. The relationship between site condition and circulation of nutrients in forest ecosystem (II). Moisture index as a means of evaluation of site condition. *J. Jap. For. Soc.*, **56**, 434-440.

Walton, D.W.H. 1985. Tensometer and jaw types in testing textiles for tensile strengths. *Int. Biodeterior.*, **21**, 303-302.

Walton, D.W.H. & Allsopp, D. 1977. A new test cloth for soil burial trials and other studies on cellulose decomposition. *Int. Biodeterior. Bull.*, **13**, 112-115.

Widden, P., Howson, G. & French, D.D. 1986. Use of cotton strips to relate fungal community structure to cellulose decomposition rates in the field. *Soil Biol. Biochem.*, **18**, 335-337.

***Wingfield, G.I.** 1980. Effect of asulam on cellulose decomposition in three soils. *Bull. environ. Contam. Toxicol.,* **24,** 473-476.

***Wingfield, G.I.** 1980. Effects of time of soil collection and storage on microbial decomposition of cellulose in soil. *Bull. environ. Contam. Toxicol.,* **24,** 671-675.

Wynn-Williams, D.D. 1979. Techniques used for studying terrestrial microbial ecology in the maritime Antarctic. In: *Cold tolerant microbes in spoilage and the environment,* edited by A.D. Russell & D. Fuller, 67-81. London: Academic Press.

Wynn-Williams, D.D. 1980. Seasonal fluctuations in microbial activity in antarctic moss peat. *Biol. J. Linn. Soc.,* **14,** 11-28.

Wynn-Williams, D.D. 1985. Comparative microbiology of moss-peat decomposition on the Scotia Arc and Antarctic Peninsula. In: *Antarctic nutrient cycles and food webs,* edited by W.R. Siegfried, P.R. Condy, & R.M. Laws, 204-210. Berlin: Springer.

*used weight loss, not tensile strength, as the measure of decomposition

Appendix III
Biodegradation studies at the US Army Natick Research Centre

Science and Advanced Technology Laboratory, Natick Research Development and Engineering Centre, Natick, Massachusetts, USA

Dr A M Kaplan is the principal investigator for microbiological studies at the US Army Natick Research Centre, where some basic and applied research is carried out on cellulosic materials such as textiles, clothing, or tentage, which have been subject to microbiological deterioration. The studies include chemical, physical or biological aspects.

Although the work is largely aimed at solving problems of deterioration, including aspects of the use of the soil burial method, Dr Kaplan's published work also discusses some more theoretical concepts concerning the interfacing of environment, substrate and micro-organisms (eg Kaplan 1977), along with the problems of relating phenomena observed in laboratory evaluations to actual field situations. He stresses that the environment is the dominant force in any biodegradation process, and considers that knowledge of the role of the specific environment in determining the course and sequence of events in any biodegradation process can be used to plan experimental studies so as to provide valid predictions of responses in the field.

Readers may like to know of the work of this group in the USA, and the following list gives some of the relevant publications.

References

Darby, R.T. & Kempton, A.G. 1962. Soil burial of fabrics treated with minimal concentrations of fungicides. *Text. Res. J.*, **32**, 548-552.

Kempton, A.C., Maisel, H. & Kaplan A.M. 1963. A study of the deterioration of fungicide-treated fabrics in soil burial. *Text. Res. J.*, **33**, 87-93.

Kaplan, A.M. 1977. Microbial degradation of materials in laboratory and natural environments. *Develop. ind. Microbiol.*, **18**, 203-211.

Kaplan, A.M. 1979. Prediction from laboratory studies of biodegradation of pollutants in 'natural' environments. In: *Proceedings of the workshop; microbial degradation of pollutants in marine environments*, edited by A.W. Bourquin & P.H. Pritchard, 479-484. (PB-298 254.) Florida.

Kaplan, A.M. 1984. Military needs and criteria for industrial biocides. *Develop. ind. Microbiol.*, **25**, 373-378.

Kaplan, A.M. & Rubidge, T. 1978. Letters to the Editor. *Int. Biodeterior. Bull.*, **14**, 31-34.

Kaplan, A.M., Mandels, M., Pillion, E. & Greenberger, M. 1970. Resistance of weathered cotton cellulose to cellulase action. *Appl. Microbiol.*, **20**, 85-93.

Rogers, M.R. & Kaplan, A.M. 1962. The necessity for scouring the 'standard' blue-line cotton fabric. *Text. Res. J.*, **32**, 161-162.

Wiley, B.J., Herbert, R.L., Armstrong, J. & Kaplan, A.M. 1982. The aerobiology of an environmental test chamber. *Mycologia*, **74**, 886-893.

Wendt, T.M., Kaplan, A.M. & Greenberger, M. 1970. Weight loss as a method for estimating the microbial deterioration of PVC film in soil burial. *Int. Biodeterior. Bull.*, **6**, 139-143.

Appendix IV
List of participants

Mr A R M BARR, Catomance Ltd, 96 Bridge Road East, Welwyn Garden City, Hertfordshire, AL7 1JW.

Mr G BANCROFT, Aynsome Laboratories Ltd, Kentsford Road, Grange-over-Sands, Cumbria, LA11 7BA.

Miss H COLLINS, Department of Soil Science, University of Aberdeen, Aberdeen, AB9 2UE, Scotland.

Mr J B DIXON, Department of Applied Biology, University of Wales Institute of Science and Technology, Llysdinam Field Centre, Newbridge-on-Wye, Llandrindod Wells, Powys, Wales.

Miss T DOYLE, School of Botany, Trinity College, Dublin 2, Ireland.

Mr J GILLESPIE, Department of Chemical and Process Engineering, University of Strathclyde, James Weir Building, 75 Montrose Street, Glasgow, G1 1XJ, Scotland.

Mr J A HARRIS, Environment & Industry Research Unit, Biology & Biochemistry Department, North East London Polytechnic, Romford Road, London, E15 4LZ.

Dr P HOLTER, Institute of Population Biology, Copenhagen University, University Park 15, DK2100, Copenhagen, Denmark.

*Professor K KUZNIAR, Akademia Rolnicza, Instytut Meliotacji Rolnych i Lesnych, Aleje Mickiewicza 24/28, 30-059 Krakow, Poland.

*Miss A LIGGETT, 22 Selworthy Green, Liverpool, L16 9JJ.

Professor A MACFADYEN, Department of Biology, University of Ulster, Coleraine, Co. Londonderry, BT52 1SA, Northern Ireland.

*Dr E MALTBY, Department of Geography, University of Exeter, Amory Building, Rennes Drive, Exeter, EX4 4RJ.

Dr J M MAW, School of Biological Sciences, Hatfield Polytechnic, PO Box 109, College Lane, Hatfield, Hertfordshire, AL10 9AB.

Dr C NYS, Sols Forestiers, Centre de Recherche Forestiere, Champenoux, 54280 Siechamps, France.

*Dr J PROCTOR, Department of Biological Sciences, University of Stirling, Stirling, FK9 4LA, Scotland.

Dr B SAGAR, Shirley Institute, Didsbury, Manchester, M20 8RX.

Dr R N SMITH, School of Natural Sciences, Hatfield Polytechnic, PO Box 109, College Lane, Hatfield, Hertfordshire, AL10 9AB.

*Professor K VAN CLEVE, Forest Soils Laboratory, O'Neill Building, University of Alaska, Fairbanks 99701, Alaska, USA.

*Dr P J VICKERY, CSIRO, Pastoral Research Laboratory, Division of Animal Production, Armidale, New South Wales 2350, Australia.

*Professor Y WAISEL, Faculty of Life Sciences, Department of Botany, Tel-Aviv University, Ramat-Aviv, Israel.

Dr D W H WALTON, British Antarctic Survey, High Cross, Madingley Road, Cambridge, CB3 OET.

*Dr P WIDDEN, Department of Biology, Concordia University, 1455 de Maisonneuse West, Montreal, Quebec, H3G 1M8, Canada.

Miss S WIGFULL, Environment & Industry Research Unit, Biology & Biochemistry Department, North East London Polytechnic, Romford Road, London, E15 4LZ.

Dr D D WYNN-WILLIAMS, British Antarctic Survey, High Cross, Madingley Road, Cambridge, CB3 OET.

Mr G XU, Institute of Forestry & Soil Science, Shenyang, People's Republic of China. (Visitor at Merlewood.)

Institute of Terrestrial Ecology staff
Merlewood Research Station, Grange-over-Sands, Cumbria, LA11 6JU

Mrs L Ashburner, Dr P J Bacon, Mr A H F Brown, Dr J Dighton, Dr J C Frankland, Dr A F Harrison, Mrs D M Howard, Mr P J A Howard, Mrs G Howson, Dr P Ineson, Miss P M Latter, Mr G J Lawson, Mr D K Lindley, Mr C Mylechraine, Mrs J P Poskitt, Mrs F J Shaw

Banchory Research Station, Hill of Brathens, Glassel, Banchory, Kincardineshire, AB3 4BY, Scotland

Mr D D French, * Dr J Miles

Monks Wood Experimental Station, Abbots Ripton, Huntingdon, PE17 2LS

Dr M O Hill

*indicates those contributing papers, or data sets for the workshop session, but not present at the Symposium

Epilogue

'Though analogy is often misleading, it is the least misleading thing we have'   (Samuel Butler)

Apart from a small workshop meeting at the University of Aston in 1977, this Symposium provided the first opportunity for a major group of cotton strip users to pool their expertise and data. What has it achieved?

The accounts in this volume represent studies at 3 levels: global, as in the comparison of world-wide sites; local, in the study of habitats, treatments and management; and micro-, as an experimental tool for microcosms and microhabitats. They reflect the uses and applications of the assay over the last 2 decades, and each gives scope for further development. From the wide range of papers and the animated discussions which took place between participants, some important general points have emerged.

The tendency to use the assay as a general index of decomposition is apparent in several of the papers. As an index of decomposition, reflecting the overall effect of the environment on one analogue substrate over a period of time, it will always have its value and its limitations, and both are critically discussed.

Our understanding of all the soil processes involved in the reduction of tensile strength is still incomplete, and the continuing discussion on the validity of analogue substrates in general is both useful and necessary. Despite this shortcoming, use of the assay in a comparative, rather than absolute, mode in which it can detect (with remarkable sensitivity) differences in the potential for decomposition in widely dispersed and/or diverse habitats is a most valuable tool to the ecologist. The papers generate some interesting concepts and ideas to be followed up, and report some intriguing patterns of activity which have already provided, or will provide, triggers to stimulate further research into the real processes underlying the differences observed in the field.

Future improvements to the interpretation of cotton strip assay data must include investigations of:

— colonization rates of the fabric in soil;
— the selectivity of cotton as a substrate;
— characterization of the succession of colonizing organisms;
— relationships with other decomposition parameters, with specific environmental factors and

with plant growth. Routine measurement of temperature is recommended using the thermal cell method (see p. 99).

Suggestions were made for impregnating strips with tannins, to mimic leaf litter, or with nutrients, to extend their use as a test material.

Given all the advantages, it might reasonably have been expected that the analysis of many data sets on a global scale would have taken our general understanding of this assay a stage forward. In fact, the examination of 329 insertions has emphasized the sparsity of the ancillary site data which are clearly essential for further progress. Even the geographical distribution of field sites for which data were available at the time of the workshop was too limited to derive any convincing global conclusions, but we have identified the important gaps in our knowledge, which fact alone justifies the present attempt at global analysis. The global synthesis has suggested that, amongst those variables tested so far, potential evapotranspiration is the best predictor of cotton rotting rate. The data base now established at the Institute of Terrestrial Ecology's Merlewood Research Station provides us with the opportunity to extend the synthesis exercise, by the addition of necessary micro-environmental data. We hope that users will continue to participate in this exercise and that a better synthesis can be published in due course.

Can we now look forward to an increasing use of the assay by ecologists world-wide, especially for comparative investigations between and within individual habitats? To those ecologists who have used the assay, its strengths are still those described over 20 years ago: cheapness; ease of replication and use by unskilled workers in the field; adaptability to various situations. The availability of a standard cloth from a single source has reduced experimental variance and increased confidence in the assay.

We hope those applying the assay, or prospective users, will gain the information they require from this volume, and that future studies will be better planned and statistically analysed as a result, leading to an international standardization of approach.